Use R!

Series Editors:
Robert Gentleman   Kurt Hornik   Giovanni Parmigiani

For further volumes:
http://www.springer.com/series/6991

Nathan G. Swenson

Functional and Phylogenetic Ecology in R

 Springer

Nathan G. Swenson
Assistant Professor of Plant Biology
Michigan State University
East Lansing, MI, USA

ISSN 2197-5736   ISSN 2197-5744 (electronic)
ISBN 978-1-4614-9541-3   ISBN 978-1-4614-9542-0 (eBook)
DOI 10.1007/978-1-4614-9542-0
Springer New York Heidelberg Dordrecht London

Library of Congress Control Number: 2013958447

Printed on acid-free paper

Springer is part of Springer Science+Business Media (www.springer.com)

To Eowyn

Preface

The hypotheses that have been derived to explain the distribution, abundance, dynamics, and diversity of species are typically tested using data matrices containing the names and abundances of species in one to many sites. In effect, this research approach treats all species as independent entities. Species, of course, are not independent. They vary in their degree of relatedness and their functional similarity. Aside from being a basic statistical "issue to deal with," not accounting for the phylogenetic and functional nonindependence of species likely severely limits the ability of an ecologist to adequately test the hypotheses of interest. Indeed the more we analyze the phylogenetic and functional signature in ecological data, the more we realize that the inferences we might have made using only information about species names and abundances would be overturned if we also considered their phylogenetic and functional similarity. I have now seen such instances enough to be wary of any ecological analyses that do not consider where the species came from and how they actually function.

As a young field ecologist working in the rainforest in El Yunque, Puerto Rico, right after earning my undergraduate degree, I was consistently interested in the spatial distributions of individual species. I could tell you where in the forest you are likely to find a certain species and whether or not other species from that same genus or family were also likely to be in that same area. This was satisfying to some degree, but I was always frustrated by not being able to tell someone *why* a species was only found along streams or ridges. My frustrations lead me to seek out institutions where I could perform doctoral research that would investigate the functional ecology of tropical trees. While interviewing for graduate school I was shocked when three potential Ph.D. mentors in a row asked me whether I was familiar with Cam Webb's research into the phylogenetic structure of tropical tree communities. Admittedly I had not read Cam's work prior to my first interview, but by the time the second and third interviews came around later that month, I had read almost all of his papers. My master's research had been focused on evolutionary biology and I was immediately excited to see phylogenetic information being woven into tropical tree ecology. Around this time I was also reading reviews and empirical studies focusing on the evolution of plant function. Suffice it to say that by the time I started my Ph.D. research that fall at the University of Arizona, I was very excited about the prospect of quantifying function in tropical tree communities and putting that information into a phylogenetic context.

Almost immediately upon arrival in Tucson, I began to analyze the phylogenetic and functional composition of tree assemblages in the Neotropics. Immediately I ran into significant computational challenges both in the form of integrating and formatting enormous forest plot datasets and in the form of synthesizing the output. I had heard of R during my master's research and it was clear many of my new lab mates were using R, but I had yet to dive in and learn. It was at this time when a postdoctoral fellow and a friend of mine named Jason Pither taught me enough basic R and gave me enough books to make me dangerous (and not always in a good way). It is hard to thank Jason enough for his initial help and encouragement. My life as a researcher has never been the same and I still consult him from time to time for sage advice.

Several years have passed now and I have transitioned slowly from a young field ecologist still trying to figure out basic R code to someone leading R workshops around the world focusing on phylogenetic and functional analyses of ecological data. The first of these workshops I conducted at the Chinese Academy of Sciences, Institute of Botany in Beijing. In preparing for that course I generated a short workbook. I continually updated and expanded the workbook prior to new workshops over the years and the result is this book. I am indebted to the students in those workshops for serving as "guinea pigs" and for helping me improve this text.

There are many additional people that I have to thank. I would first like to thank Xiangcheng Mi for being a great collaborator over the past several years where we have both benefitted from the R expertise of the other. I would also like to thank my other close colleagues at the Institute of Botany in Beijing (Keping Ma, Jinlong Zhang, and Xiaojuan Liu) and at the Xishuangbanna Tropical Botanical Garden (Min Cao, Luxiang Lin, and Jie Yang). We have all shared many fun collaborative experiences and all share a zeal for tropical trees that keeps us going. I wish to thank the researchers affiliated with the Smithsonian Institution Center for Tropical Forest Science that have always been generous in their collaboration. In particular, I thank Stuart Davies, Rick Condit, John Kress, Dave Erickson, Joe Wright, Jess Zimmerman, Jill Thompson, Liza Comita, Andy Jones, Nathan Kraft, Bob Howe, and Amy Wolf. I would also like to thank the many researchers who have generated the R code constituting the packages that are utilized in this book. Writing these packages has been a huge service to the research community. I would like to thank Springer and the Editors of the UseR! series for encouraging me to contribute this text. I would like to thank the United States National Science Foundation for generously funding my phylogenetic and functional ecology research from my postdoctoral research to the present. Lastly, I would like to thank Liwei Hua for all of her love and support before, during and after I composed this book.

Aarhus, Denmark   Nate Swenson

Contents

**9 Integrating R with Other Phylogenetic and Functional
Trait Analytical Software** 189

Chapter 1
Introduction

1.1 Why Phylogenetics and Functional Traits in Ecology?

The distribution of biodiversity is a, if not the, major focus of ecologists. Specifically, ecologists often investigate the spatial or temporal trends in biodiversity levels within a particular study region or across the planet. The study of biodiversity has traditionally focused on quantifying patterns of species diversity or species richness across some type of gradient and determining the potential processes that have produced the observed pattern. This approach is a cornerstone of ecological investigations and thinking regarding biodiversity. However, there are two clear limitations to this species-centric approach. First, biodiversity is not simply species diversity. Biodiversity also includes the phylogenetic, genetic, and functional diversity in an assemblage [1]. Indeed, species diversity may even be the least informative of all of these dimensions of biodiversity. For example, regions could have the same exact species diversity, but very different levels of phylogenetic and functional diversity and therefore very different levels of biodiversity. Or they could have very similar levels of functional and phylogenetic diversity despite large differences in their species richness [2–5]. Thus, attempting to determine the processes that produce biodiversity cannot be obtained by examining only one component of biodiversity. A second challenge for the species-centric approach to studying biodiversity that is perhaps more important than the first one is that species names are relatively information poor. While they are fundamental to biology, they convey little information regarding the function or evolutionary history of species, and such information is critical for determining the processes that have combined to produce the observed levels of biodiversity. These inherent limitations of a species-centric approach suggest that a more pluralistic approach to studying biodiversity is needed in order to obtain a mechanistic understanding of how patterns of biodiversity are formed [6–13]. In particular, a biodiversity synthesis will necessarily require the consideration of the interrelationships between the three primary components of biodiversity—species diversity, functional trait diversity, and phylogenetic diversity [1]. Ecologists are now embracing this reality and have altered their research programs accordingly.

N.G. Swenson, *Functional and Phylogenetic Ecology in R*, Use R!,
DOI 10.1007/978-1-4614-9542-0_1, © Springer Science+Business Media New York 2014

The number of phylogenetic- and functional trait-based analyses in ecology has skyrocketed in recent years resulting in hundreds of publications. Indeed, entirely new fields in ecology have formed such as community phylogenetics, and new grant programs have sprung up such as the United States National Science Foundation's Dimensions of Biodiversity program.

Coinciding with the increased interest in quantifying phylogenetic and functional diversity in ecology, a dizzying array of tools and methods has been generated to incorporate phylogenetic and functional information into traditional ecological analyses. Increasingly, these tools are being implemented in R making them easily and freely accessible to researchers around the planet. The goal of this volume is to lead beginning or advanced R users through phylogenetic- and functional trait-based ecological analyses in R. It is expected that beginning users can use this volume as a step-by-step entryway into phylogenetic and functional analyses for ecology in R, whereas it is expected that more advanced users will be able to use this volume as a "cookbook" or quick reference to understand particular analyses. The volume starts with chapters on the R environment and phylogenetic data in R. These are followed by three chapters providing comprehensive coverage of phylogenetic and functional metrics of biodiversity and one chapter on null modeling and randomizations for phylogenetic and functional trait analyses in R. Lastly, two chapters focusing on integrating phylogenetic and functional trait information are provided followed by a final chapter that focuses on interfacing the R environment with a commonly used C-based program called Phylocom that has been influential in phylogenetic ecology [14].

1.2 Why R?

After learning how to ask fundamentally important questions, basic natural history and field identification of the organisms in their study system, I think there are few skills more useful for young ecologists to learn than programming in general and statistical programming in particular. Ecology, like many other disciplines, is rapidly advancing in its analytical complexity and its utilization of "big data." Performing advanced analyses, even on small datasets, or performing even simple analyses on large datasets typically require some level of comfort with computer code. Indeed, when I meet ecology undergraduate and graduate students (and faculty) at other universities I am often asked whether it is "worth it" to learn a programming language like R. I usually provide the response "of course." In many cases I am met with an unconvinced look. I can read in their eyes that they really don't buy my response as a reason to go through what seems to be a daunting process of learning a computer language. To combat this response I often like to first say learning R is very liberating as it frees one up to do many more analyses that they can currently perform. Second, to buoy the first statement I convey an estimate of the percent of the journal articles that I have published that I think would have been possible without R or the ability to program in some other language. The percent I generally estimate is surprisingly small to many (<20 %). In other words, a lot of the work I do simply would never be possible without even a basic ability to write computer code. I was fortunate enough to be confronted with this reality very early on in my graduate career while working with big datasets and I realized that I better learn a programming language quick if I was to finish my Ph.D. in under a decade.

Learning R and using it day to day (when not in the field, but often also using it in the field letting analyses run on my computer in the field station while I was out staring at trees) was perhaps one of the most valuable tools I gained in graduate school. While it is certainly valuable to learn other programming languages, I would argue learning R is the best starting place for ecologists. This is because ecology and many other disciplines have converged on R for their statistical analyses. This creates a positive feedback loop where more and more researchers perform their analyses in R and write analytical code specifically for R, and therefore more researchers find themselves drawn into the R universe and also contribute. The R code that researchers produce is often made available in packages or in the supplemental material of journal articles making analyses transparent and widely accessible. The issue of accessibility brings us to another important reason why R should be used. R is free! You do not have to pay large sums of money to run your statistics and neither do your collaborators with whom you would like to share your code. Anyone anywhere can freely download the software on their computer and run the most current and advanced analytical code in their field. This greatly levels the analytical playing field for ecology and that can only be a good thing for our science and for achieving our common goals. So I ask you—why not R? The most advanced ecological analyses are now generally coded in R, and this code is becoming or already has become the common analytical currency in ecology.

1.3   Structure and How to Use This Book?

The book is designed to introduce you to phylogenetic and functional trait analyses that can be performed in R. I will not describe in detail each chapter here, but if you are new to R and/or new to phylogenies you should not skip past Chap. 2. This chapter introduces you to phylogenetic data in R—how it is structured and how it can be plotted and manipulated to meet your research goals. This chapter is simply designed as a primer for ecologists and not a comprehensive treatment on phylogenetics in R. At the given time there are enough R packages for phylogenetic analyses that such a treatment would be hard to compile, but I do highly recommend *Analysis of Phylogenetics and Evolution in R* by Emmanuel Paradis in the Springer UseR! series [15] as a wonderful introduction to phylogenies in R and comparative analyses. I would also highly recommend reading a few key texts regarding phylogenetics and comparative methods to help you fully understand what goes into inferring phylogenies and analyzing data in a phylogenetic context (e.g., [16–18]). The present book will cover some similar topics covered in the Paradis book [15] related to how to handle and plot phylogenetic data in Chap. 2 and comparative analyses in Chap. 7, but the result of the present book is a significant departure focusing primarily on ecological analyses and not macroevolutionary analyses per se. Similarly, the UseR! series book *Numerical Ecology in R* by Daniel Bocard, Francois Gillet, and Pierre Legendre [19] would likely be of interest and use to the readers of this book for general ecological analyses, but the present book significantly differs due to its exclusive focus on phylogenetic and trait data.

The vast majority of the analyses to be discussed in this book can be accomplished using simple "plug and chug" functions in a variety of existing R packages. In teaching courses and workshops on these topics, I have come to two main conclusions. The first is that the participants in my courses and workshops often find it very difficult to navigate the large number of packages available, and they find it difficult to determine whether certain functions do what they want or are similar to other functions in other packages. The second is that participants in my courses and workshops can very easily type in a line of code and get a result, but learning to do this is not very beneficial by itself. This is because the student doesn't realize exactly what was calculated and how it was done. This causes serious problems and limitations when the time comes to interpret the results or when a researcher decides they would like to modify the analytical approach to suit their particular needs. With these issues in mind, the majority of the code provided in the book is designed to lead you, the reader, through the computational steps necessary to calculate the metrics being discussed. I will use my own code to achieve this goal. In some cases my code will be very similar to that in the "canned" R functions already available, and in other cases the code may be significantly different but produces the same result. This difference can be due to different coding styles between me and the original author or my attempts to speed up code by using functions like `apply()` instead of `for()`. After we have broken an analysis into its individual components and calculated the desired result, I will provide you with the name of the "canned" function in an R package, where possible, that should provide the same result. Thus, if you wish to eschew learning how the fine details regarding how the analysis works and what it means, you can ultimately just use the functions highlighted at the end of each section. Though, I don't recommend this approach and I hope that you read, work, and think through the components of the code I provide so you have a detailed understanding of how that number you receive at the end was calculated and what it means. By doing this you will also learn R and learn how to tinker with R code for phylogenetic and functional trait analyses so that you can customize new analyses for your own particular dataset. While working through the examples in the book, you may find that you often run across a new function that you were not aware of before and you may want a more detailed description of what is provided to that function and what comes out of it. To get this information you can access the help file for any function using a "?" and then the function name. For example, if you wanted to know what the `mean()` function does in R, you could see the help page for this function by typing:

```
> ?mean
```

I highly recommend taking this approach and I often do it myself to navigate the code of others and to remind myself how to use a function that I use infrequently.

At the end of each chapter you will find a series of exercises. Some of the exercises will be quite simple and are simply designed to get you used to running the analysis on a variety of datasets. Other exercises will require you recall the information you learned in previous chapters. The goal of this is to help you integrate concepts and information and to help you memorize code and analyses through repetition in different venues. A few exercises will require you to use functions we have not covered. These exercises will be much more difficult to accomplish for the new user, but I have generally told you what new functions you will likely have to use to accomplish the task. It is then up to you to discover how to use these functions and put them together to solve the problem. This is the type of practical problem you will encounter in your future work where you have a particular problem to solve; you break that large problem into many small problems that can be solved with the right tools (i.e., R functions) and then you integrate the solutions to all of those problems to solve the one large problem. Going through advanced exercises such as these will rapidly help you become a more powerful R user.

Lastly, the book relies on many example datasets. Some of these datasets are subsets of larger datasets I utilize for my own work in plant ecology. Others are datasets I have "cooked up" in R. I encourage you to first utilize these datasets to run the analyses, but you should then quickly transition to using your own datasets. As you will quickly find out it is very easy to plug and chug with example datasets, but tiny problems will lurk when you use your own dataset. While these issues regarding minor differences in formatting between files, for example, may be frustrating, it is a common obstacle in data analysis and learning to confront these problems sooner rather than later will be useful to you.

1.4   Setting Working Directories and Package Installation

This book is intended for an audience that spans researchers that are relatively new to R to more advanced R users all of whom would like to incorporate phylogenetic and/or functional information into their research programs. In order to span this gradient, it is necessary to cover some basics that an advanced user does not need to review. For those advanced users this subsection will not be that useful aside from the list of packages at the end that we will utilize in this book. Relatively, new R users may need this section for a brief review on what working directories are and how they are set and how R packages are installed and loaded. We will begin with discussing working directories.

The working directory is the folder (a.k.a. directory) on your hard drive in which the files that you are using and creating are stored. For example, you have a file for your phylogenetic tree and a file for your community data in a folder and you would like to read those files into R and generate output in R that you can write to this same folder.

This can be accomplished by typing in the path to your files every time you read and write them into and out of R, but it is often easier to simply set a single working directory for the project you are working on at that time. For those of you still not totally excited about typing in commands, you can set the working directory in R using drop down menus:

For PC it is under the "File" menu as "Change dir…"
For MAC it is under the "MISC" menu as "Change Working Directory…"

For those of you ready to take the plunge and start typing commands, you can set working directories if you know the "path" to your file as. For example:

For PC:

```
> setwd("C:/my.working.directory/")
```

For Mac:

```
> setwd("/Users/Nate/my.R.project.folder")
```

You can always find out your current working directory by typing:

```
> getwd()
```

Getting the current working directory can be a good way to find out "where you are" currently so you can set a new path for your desired working directory. If you are using the drop down menu to set your working directory, it is useful to get the current working directory path using `getwd()` so you can begin to learn what a path looks like and how to define one. Once you have set your working directory you can obtain a list of all of the files in that working directory. You can accomplish this for any directory, actually, on your computer, but for your current working directory you can simply type:

```
> list.files(getwd())
```

You will now see a list of file names print out on your R console. These are all of the files currently contained in your working directory.

We are almost ready to jump right in and proceed to the next chapter on phylogenetic data in R, but first we must discuss packages. R packages contain a series of functions that can be used for data manipulation or analysis. The functions in a package may use each other and often other functions written in other packages. That is, a function X in package A may need to perform a sub-analysis that can be performed by function Y in package B. In those instances function X in package A "depends" on function Y in package B. Such dependencies are commonplace and one of the wonderful things about R such that you don't have to write your own function anew for your package. You can simply call a function from another package. In this book we will use many packages for phylogenetic and trait analyses in ecology that are useful by themselves or integrated with other packages.

To install a single package, in this example the R package *vegan*, you can type the following command:

```
> install.packages("vegan", dependencies = TRUE)
```

After typing in this command and hitting return you may be asked to select a "mirror." Simply select a mirror that is closest to your geographic location. This will then be the default location from where you download your desired R packages for the current session. Once you have downloaded an R package, it is there for each R session in the future, but its functions are not available to you immediately until you load the package into memory. To load an R package at the start of your session or midway through a session when you need a new function from a different package, you can use the following code again using *vegan* as our example:

```
> library(vegan)
```

If the package you are loading depends on other packages that are already downloaded on your computer, the other packages will also be loaded. If those packages necessary are not on your computer you will receive a message telling you that they must be downloaded.

Now that we have covered how to install and load specific packages, I will simply list the packages that we will use in this book. These could be all installed en masse, but since you may not want to utilize all analyses in this book or you may not want these many items downloaded to your hard drive I will only list them and expect that you can download those that you need or want when you see me call the library in the code in a chapter. The specific packages we will use are: *ape, vegan, phytools, geiger, abind, picante, Rsundials, nlme, adephylo, phylobase, ecodist, ade4, bipartite, geometry, packfor, GUniFrac, SDMtools, fBasics,* and *FD*.

Chapter 2
Phylogenetic Data in R

2.1 Objectives

The objectives of this chapter are to introduce the user to how phylogenetic information is stored, presented, and manipulated in R. We will cover primarily the class "phylo" in this chapter since that is the class that is most frequently utilized in phylogenetic diversity and comparative analyses in R. By the end of this chapter the user should have a basic command of how to plot phylogenies and extract information from the data files. The chapter is designed for beginners, and users familiar with using phylogenies in R may quickly scan this chapter to refresh.

2.2 Loading Phylogenies into R and the Structure of the "Phylo" Class

This section deals with loading phylogenetic data files into R as a "phylo" class and how the data are structured inside R. The focus will be primarily on learning the basics nuts and bolts of phylogenetic data in R. As such there will be a number of aspects of phylogenetic data that are not germane to the goals of this book that will not be covered. We will begin by simply reading in an example phylogeny that is in the parenthetical Newick format. This can be accomplished using the `read.tree()` function in the *ape* R package. Thus, we must first load the *ape* package and then use the function.

```
> library(ape)

> my.phylo <- read.tree("my.phylo.newick.file.txt")
```

The online version of this chapter (doi: 10.1007/978-1-4614-9542-0_2) contains supplementary material, which is available to authorized users

N.G. Swenson, *Functional and Phylogenetic Ecology in R*, Use R!,
DOI 10.1007/978-1-4614-9542-0_2, © Springer Science+Business Media New York 2014

To print the original file to your R console so that you can see the original Newick format simply use the `write.tree()` function.

```
> write.tree(my.phylo)
```

Alternatively, you could also open the original file in a text editor. We can also read a Nexus formatted phylogeny into R using the `read.nexus()` function in the *ape* package.

```
> my.nexus.phylo <-
read.nexus("my.phylo.nexus.file.txt")
```

To print the original Nexus format to your R console so that you can see the original Nexus format simply use the `write.nexus()` function.

```
> write.nexus(my.phylo)
```

The Nexus file could also be opened with a text editor or with the program Mesquite [20]. Both files are now stored as class "phylo" objects in R. Thus, the two files that were once in different formats are now stored using a structure in R allowing us to use either version in the following analyses in this chapter and the majority of this book. While Nexus files can and often do hold much more information along with the phylogenetic tree such as trait data and sequence alignments, in this instance the Nexus file does not hold additional information and we will use the Newick version stored in the `my.phylo` object for the remaining analyses.

Now that our phylogeny is in R we should explore its contents and structure first before we do any analyses just like you should do for any other data file you would have in R. To do this we can first simply type the name of the object.

```
> my.phylo
```

We see that our example phylogeny contains 26 tips and 25 internal nodes. The phylogeny also has "terminal nodes," which are the tips. In some cases a phylogenetic tree may not be fully bifurcating. That is, a single branch may not split into two branches at an internal node and may branch simultaneously into three or more branches. This is called a "polytomy." A polytomy can be further classified as a "hard polytomy" where the branching of three or more lineages from one is to signify an actual simultaneous divergence or a "soft polytomy" where the branch of three or more lineages from one is used to signify ignorance regarding who is most closely related to whom between the derived and ancestral lineages. Typically, the polytomies that you will encounter will be soft polytomies. A quick way to tell whether your phylogeny probably contains a polytomy is that the number of internal nodes is less than the number of tips minus 1.

The next item printed out by R is a series of tip labels for the first six tips. The next item lists node labels if they exist for our phylogeny. In many cases you will not have node labels in your phylogeny. Finally, we see that the phylogeny is rooted and has branch lengths. This is all useful information that we can quickly glean, but it is simply printed on our screen and is not useful for more detailed investigations into our phylogeny. In some cases we can simply ask R whether our phylogeny has a particular structure. For example, we can ask if the phylogeny is rooted or if it is ultrametric (i.e., all tips end at the same point as is true for a phylogeny of extant species scaled to time).

```
> is.rooted(my.phylo)

> is.ultrametric(my.phylo)
```

Though, even the answers to these questions are not only a fraction of what we might want to know or extract from our phylogenetic tree in R and we therefore need to learn more about how the phylogenetic data is structured in R. To learn more about how our phylogenetic data is structured, we can ask for the names of our phylogeny object.

```
> names(my.phylo)
```

We see that five names: "edge," "tip.label," "Nnode," "node.label," and "edge.length" are reported. We can extract each of these from the object using the $ symbol and the name. For example, to examine what the "edge" is we can do the following:

```
> my.phylo$edge
```

This results in a matrix with two columns, and the number of rows corresponds to the number of branches in our phylogeny. The values in the first column correspond to the internal node from where the branch originates, and the values in the second column correspond to the internal or terminal node where the branch ends. The next name we can examine is "tip.label."

```
> my.phylo$tip.label
```

This returned a vector containing all of the names of the taxa on the tips of our phylogeny. We will examine this later, but the order of the names in this vector is from the species on the bottom of the phylogeny when plotted using `plot(my.phylo)`, and the last name is the species on the top of the phylogeny when plotted using `plot(my.phylo)`. We now examine what our phylogeny object holds under the name "Nnode."

```
> my.phylo$Nnode
```

The number reported is simply the number of internal nodes in our phylogeny. Next we can ask what is contained under the "node.label" name.

```
> my.phylo$node.label
```

This produces a vector of the internal node labels in our phylogeny with one label per node. In this instance the node labels are simply the numbers 1 through 25. Recall, that many phylogenies will not have node labels or may have labels for only some of the internal nodes. Lastly, we can ask what is under the "edge.length" name in our object.

```
> my.phylo$edge.length
```

A vector of branch (i.e., edge) lengths is returned. The length of this vector is equal to the number of branches in our phylogeny, and the order of the values corresponds to the order of the branches described under `my.phylo$edge`. Thus, if we wanted to make a matrix that had the information regarding the node numbers for the beginning and end points of each branch in the first two columns and the length of that branch in the third column we could do the following:

```
> cbind(my.phylo$edge, my.phylo$edge.length)
```

We now have investigated the basic structure of our phylogenetic data object in R. In Sect. 2.4 we will discuss how to modify this data or to use it to calculate additional information, but first we should discuss how to plot phylogenetic trees in R because it will help us visualize downstream manipulations we will perform.

2.3 Plotting Phylogenetic Trees in R

In any situation it is always a good idea to first look at your data once it is read into R. In the case of a table or matrix one can plot the data as a histogram or just look at the numbers themselves. In the case of phylogenetic trees, a first approach can be to examine the object using the approach we used in the previous section. After this initial examination it is good to simply plot the phylogenetic tree to further investigate the data to assure there is nothing out of place or "odd" about your data. There are two easy ways to plot a phylogenetic tree in R. The first just uses the generic `plot()` function (Fig. 2.1):

```
> plot(my.phylo)
```

If we would like to decrease the size of the names of terminal taxa, in this case species, in the plot we can adjust their size by changing the `cex` value to smaller than the default of one. We can also add a scale bar using a second function called `add.scale.bar()` and providing the length desired for the scale bar (Fig. 2.2).

```
> plot(my.phylo, cex = 0.8)

> add.scale.bar( , length = 0.1)
```

The `plot()` function is useful for a quick examination, but it can be limited as it is not specialized for phylogenetic data. The function `plot.phylo()` in the *ape*

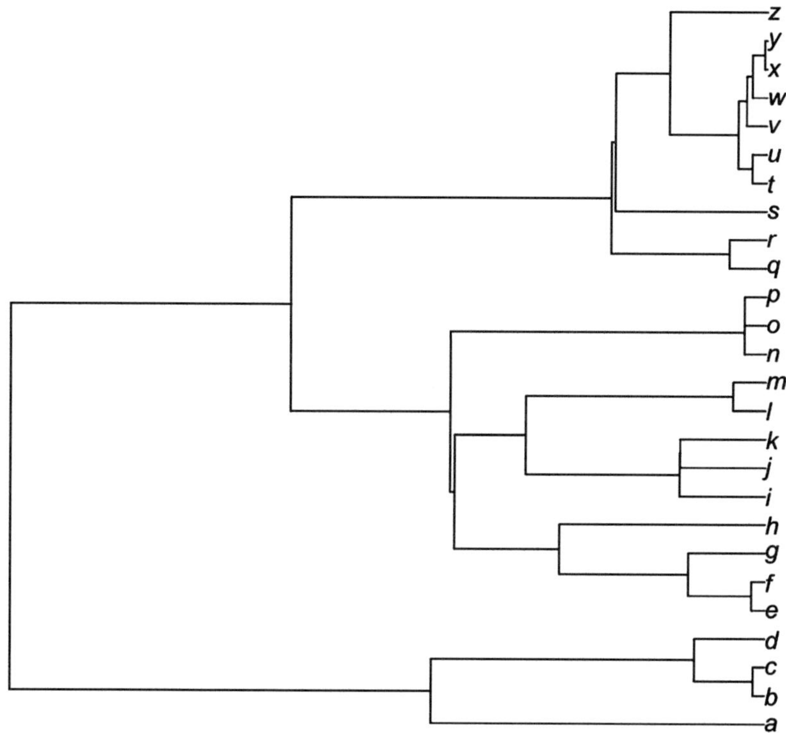

Fig. 2.1 The phylogenetic tree for our dataset

package is more flexible and designed to specifically plot phylogenetic objects. This function allows the user to display the phylogeny in different styles or types and to manipulate species names and tree branches. For example, we can plot our phylogeny as a "fan," as is often done for phylogenies much larger than ours, with gray branches (Fig. 2.3).

```
> plot.phylo(my.phylo, type = "fan", show.tip.label = TRUE,
show.node.label = FALSE, edge.color = "gray", edge.width = 1,
tip.color = "black")
```

As you can see from the code there are many options available to us regarding how to plot the phylogeny including whether or not to show labels, how to color the branches, and how to color the terminal taxa. This list does not include all of the possible ways to alter the appearance of the phylogeny using plot.phylo(). For a full description of all the possibilities, simply type ?plot.phylo.

Simply plotting your phylogeny in this way or with some other exotic format is fun and can be useful and you may find that you may want to add additional information to "decorate" your tree and this may not be possible using the plot. phylo() function. For example, in the previous section we looked at the node numbers for each branch, but we may not know how those node numbers are arrayed

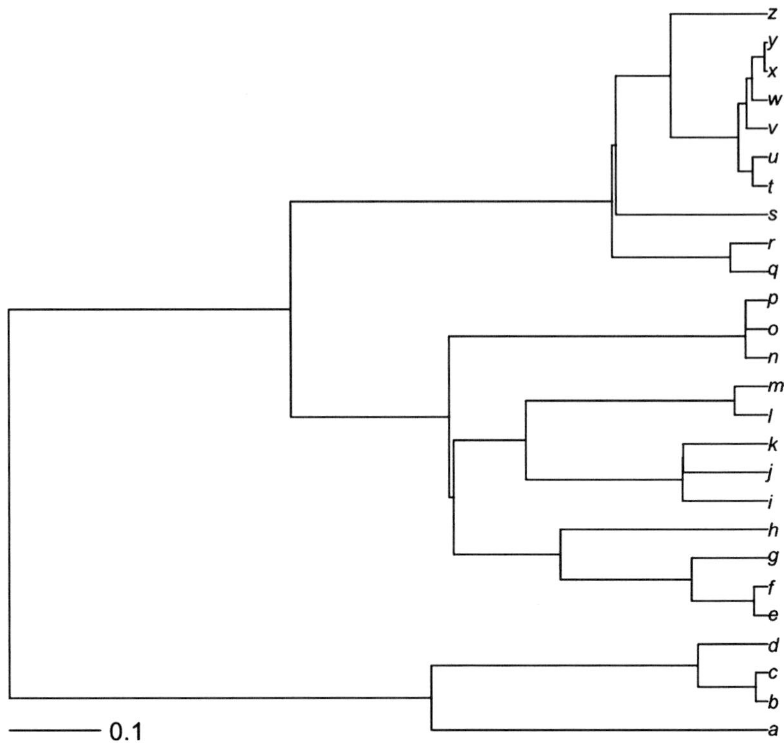

Fig. 2.2 The phylogenetic tree for our dataset with smaller labels and a branch length scale bar

on our phylogeny. To solve this problem we could first plot our phylogeny, in this example with `plot()`.

```
> plot(my.phylo)
```

Next we could simply ask R to place the node number in black for each internal node in a gray box on that node using the `nodelabels()` function (Fig. 2.4).

```
> nodelabels( , col = "black", bg = "gray")
```

You will note that the node label nearest to the root of the phylogeny is one more than the number of species in our phylogeny. That is because the first 26 node numbers correspond to the 26 species in our phylogeny. To see this you can use the `tiplabels()` function (Fig. 2.5).

```
> tiplabels(, col = "black", bg = "gray")
```

A similar procedure could be used if you had a value for each internal node sorted in the same order as the node numbers. For example, you may have an ancestral trait value estimated for each internal node that you would like to assign. This could be

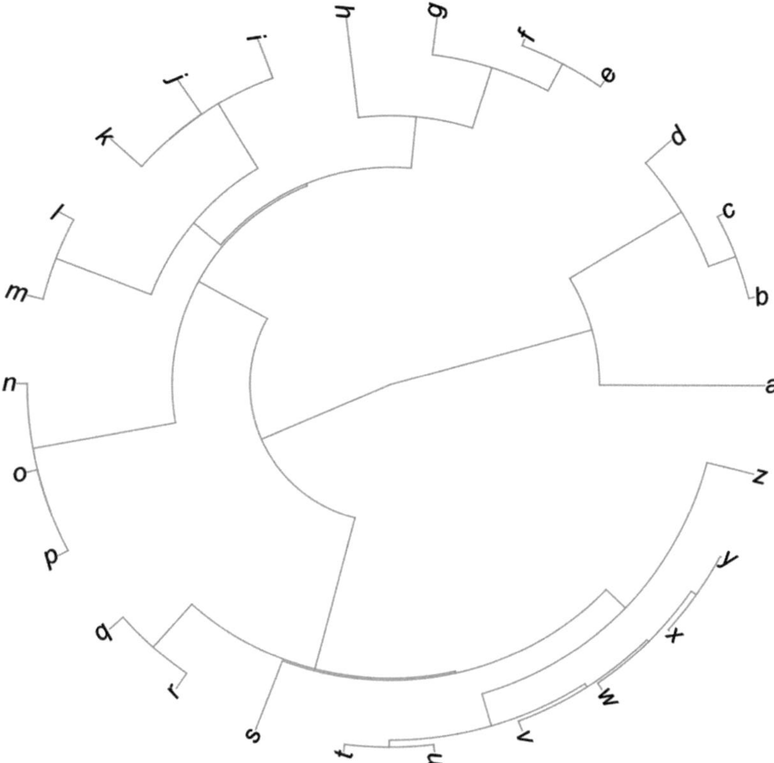

Fig. 2.3 The phylogenetic tree for our dataset displayed as a circular phylogeny or "fan" type phylogeny

done simply by providing a vector of the values to the `nodelabels()` function. There are a number of additional ways to display and decorate your tree with differing degrees of usefulness. It is not the goal to cover those more exotic approaches presently, and I will leave you to explore the wonderful diversity of ways one can plot a phylogeny in R. For the present time we can be satisfied with simply visualizing the basic structure of our phylogeny and we can proceed with how to manipulate that information or to calculate additional information from the basic structure.

2.4   Manipulating and Calculating Additional Information from Phylogenetic Trees in R

We now know how to examine the structure of our phylogenetic data objects in R and how to plot the phylogenies for visualization. The next step is to learn how to manipulate or extract additional information from the phylogenetic trees.

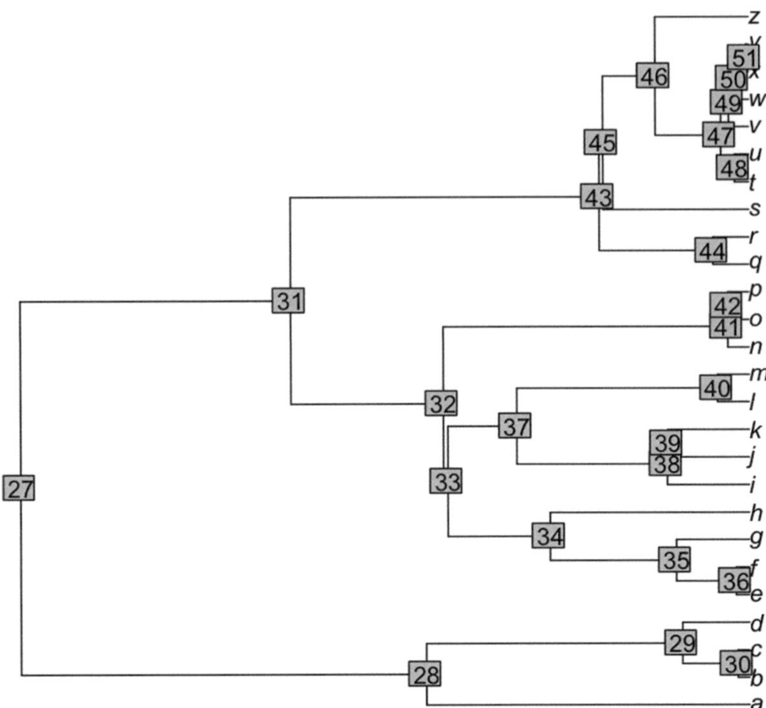

Fig. 2.4 The phylogenetic tree for our dataset with internal nodes labeled with their node numbers

We will begin by learning how to extract subsets of our original phylogenetic tree. This can be accomplished in two ways—by extracting entire clades or by pruning particular taxa out of the phylogeny. Both approaches can be useful for an ecologist. The first is useful if one wanted to perform an analysis only on a particular clade and the second is useful if one is given a phylogeny that contains more species than are in the community or trait data set being analyzed. To extract individual clades from our phylogeny, we can use the `subtrees()` function. This function takes an input phylogeny and produces a list where each element is a phylogeny object of class "phylo" containing the species derived from an individual internal node in the phylogeny. Thus, the length of the list should be equal to the number of the internal nodes in the phylogeny.

```
> my.subtrees <- subtrees(my.phylo)
```

We can look at the 15th individual subtree by asking for the 15th element in the list produced by the `subtrees()` function.

```
> my.subtrees[[15]]
```

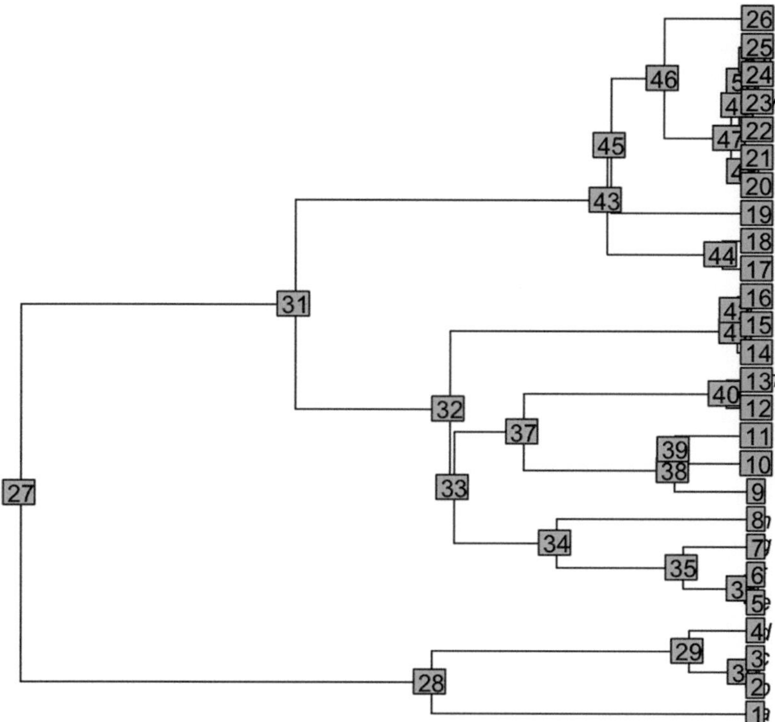

Fig. 2.5 The phylogenetic tree for our dataset with internal and terminal nodes labeled with their node numbers

We see that this particular subtree contains three taxa and two internal nodes. We can visualize this subtree by plotting it (Fig. 2.6).

```
> plot(my.subtrees[[15]])
```

The alternative to extracting all subtrees or clades from the original phylogeny is to selectively prune individual taxa out of the phylogeny. This can be done using the `drop.tip()` function, which takes a phylogeny object and a vector of names to be pruned from the phylogeny (Fig. 2.7).

```
> drop.tree <- drop.tip(my.phylo, c("e", "j", "s"))
```

```
> plot(drop.tree)
```

We can see that the three species we specified were pruned out of our original phylogeny. This approach can therefore be a powerful tool, but the downside is that you must know all of the names you don't want to keep whereas it may be easier to know all of the names you do want to keep. Later in the book we will address this situation.

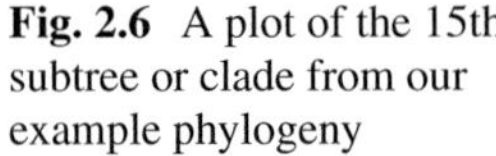

Fig. 2.6 A plot of the 15th subtree or clade from our example phylogeny

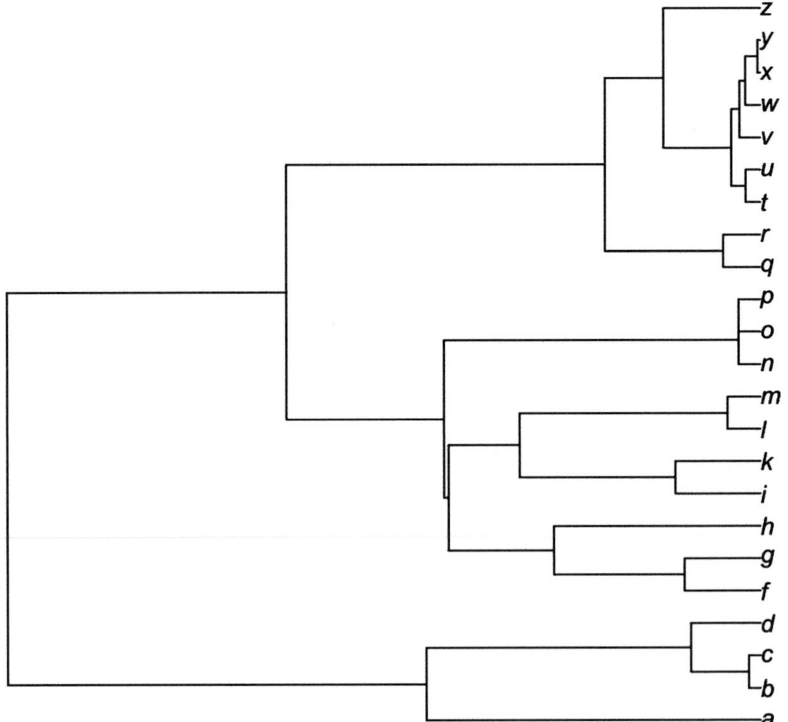

Fig. 2.7 A plot of our example phylogeny with species "e," "j," and "s" removed or pruned. Compare with Fig. 2.2 to visualize the difference

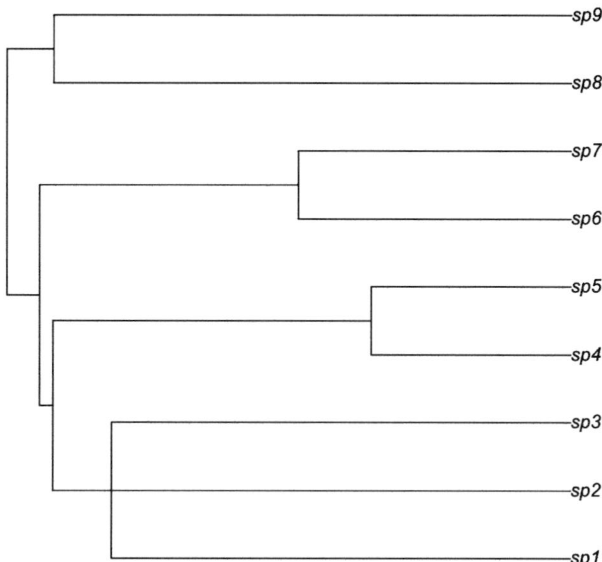

Fig. 2.8 A plot of our example phylogeny containing a single soft polytomy indicating uncertainty regarding the relatedness of sp1, sp2, and sp3

Next let us consider a situation where our phylogeny is not fully bifurcating. In many cases such a phylogeny will cause no problems for the analyses we will cover in this book, but in some cases the functions check that the phylogeny is fully bifurcating before running the analysis. If the phylogeny is not bifurcating the function will not run. Because our original example phylogeny was fully bifurcating, let us read in an example phylogeny with a single node in it that splits into three daughter lineages and not two (Fig. 2.8).

```
> my.poly.phylo <- read.tree("example.poly.txt")

> my.poly.phylo

> plot(my.poly.phylo)
```

The node with three branches emerging from a single node represents our uncertainty regarding the true relationship between these three species. There are three possibilities: the first two lineages are more closely related to each other than the third lineage, the first and third lineages are more closely related to each other than the second lineage, or the second and third lineages are more closely related to each other than the first lineage. When this uncertainty occurs it is often good practice to randomly resolve the polytomy several times and rerun the analysis each time to estimate the sensitivity of the result to the uncertainty (e.g., [21]). In the above simple case there are only three possibilities, but imagine a situation with multiple polytomies some of which are nested. The number of possibilities could be quite large. Further, when a polytomy of three lineages is resolved, a new node must be added to the phylogeny and the distance from the original polytomous node to the new node is unknown and varying this distance introduces a massive number of

Fig. 2.9 A plot of our example polytomous phylogeny where we have randomly resolved the polytomy for sp1, sp2, and sp3. Note that the phylogenetic tree still appears to contain a polytomy

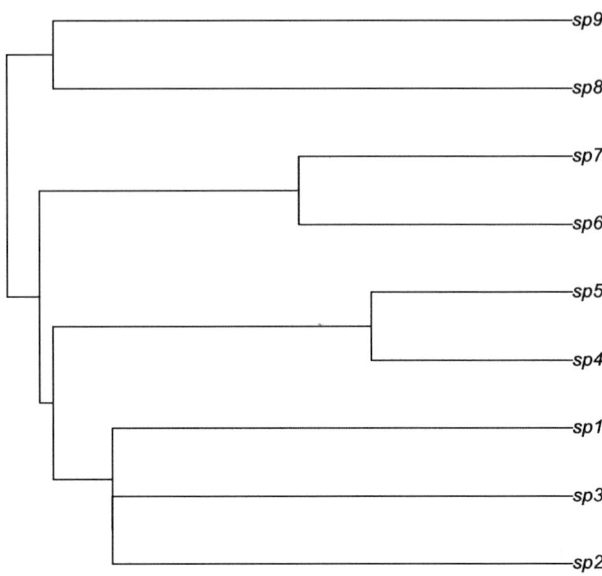

possibilities to consider. A first step that some phylogenetic programs, such as Mesquite [20], take is simply placing the new node at half the branch length from the polytomous node to its daughter node. Unfortunately, as far as I am aware, functions currently in R do not have this particular capability, but there is a function called `multi2di()` that randomly resolves polytomies in your phylogeny, but as we will see in a second the new branch lengths separating the species are zero. In this sense the `multi2di()` function tricks the R function you are trying to run into thinking the phylogeny is bifurcating, but in reality it has placed a new node at zero distance from the original polytomous node. Despite this, the `multi2di()` function can be a quick and unbiased way to manipulate your polytomous phylogeny so that the function you are interested in implementing will run.

```
> my.di.phylo <- multi2di(my.poly.phylo)

> my.di.phylo
```

We see that the new phylogeny has one more node than the original phylogeny we read into R and that the number of internal nodes is now one less than the number of tips. Though when we plot the bifurcating phylogeny we see that the tree still looks to have a polytomy. This is because the new node randomly resolving the polytomy has been placed zero distance for the original polytomous node (Fig. 2.9).

```
> plot(my.di.phylo)
```

Our next goal is to extract the branching time for each internal node for an ultrametric phylogeny. This can be done simply using the `branching.times()` function in the *ape* package.

```
> branching.times(my.phylo)
```

The output is a vector of values with a length equal to the number of internal nodes in the phylogeny. The order of the nodes is from the root of the phylogeny towards the tips. It is important to know that this function only works for an ultrametric phylogeny. If the phylogeny is not ultrametric some nodes may have a negative branching time because they are distal of some terminal taxa.

The last piece of information that we can extract from a phylogeny that we will discuss in this chapter is a matrix depicting the phylogenetic distance between each pair of terminal taxa or a matrix that depicts the amount of shared branch length between each pair of terminal taxa. The first type of matrix is typically referred to as a phylogenetic distance matrix, and it forms the foundation for many phylogenetic diversity metrics. The phylogenetic distance matrix can be generated using the `cophenetic()` function.

```
> p.dist.mat <- cophenetic(my.phylo)
```

Because the matrix is large (25×25 species) we can look at just the top left corner to get an idea of the output.

```
> p.dist.mat[1:4, 1:4]
```

We see that the resulting matrix has the species names as the row and column names, and the values in the matrix are the sum of the branch lengths separating each pair of species. Thus, if two species are far apart on the phylogeny their distance will be larger than that for two closely related species. The second type of matrix reports the shared branch length between all pairs of species and is often referred to as a phylogenetic variance–covariance (VCV) matrix. This type of matrix is used in some phylogenetic diversity metrics, but it is more commonly used in comparative analyses. A phylogenetic VCV matrix can be computed in R using the `vcv()` function.

```
> vcv(my.phylo)
```

The diagonal values of the matrix are the distances from the root to the tip that contains that species. The off diagonal values indicate the amount of shared branch length between two species. Assuming a Brownian Motion model of trait evolution, the diagonal values are used to estimate the expected variance in a trait, and the off diagonal values are used to estimate the expected covariance in the trait values between two species. I will explain this in more detail in later chapters, but essentially high off diagonal values mean species that are more closely related and are expected to have more similar trait values.

2.5   Simulating Phylogenies in R

Simulation studies are increasingly found in the ecological and evolutionary literature. Some of these studies are strictly experimental in that they experiment with different parameters to observe the likely patterns that can result. Other approaches use simulations to estimate the parameter values that would best explain an observed pattern in an empirical dataset of interest. Both are useful and now widely employed approaches in phylogenetic investigations in ecology and evolution [22–24].

The simulation of phylogenies in R is trivial, but this does not mean such simulations should be used without careful thought. There are a number of ways to simulate phylogenies in R, but we will focus on the two most basic approaches available in the *ape* package. The first approach begins with a single branch that randomly splits into two "daughter" branches. These daughter branches may then branch themselves to produce two daughters and so on until the desired number of terminal nodes is reached. This phylogenetic simulation can be performed in R using the `rtree()` function. Here, we will generate a random phylogeny with random splitting containing 40 species.

```
> new.tree <- rtree(40, rooted = TRUE, tip.label = NULL)
```

This function can be modified to include a root in the phylogeny or not. The default is to include a root. Now plot your new simulated tree (Fig. 2.10):

```
> plot(new.tree)
```

The first thing you may notice is that the phylogeny is not ultrametric and that it will not look like mine because it was randomly generated. You may also notice that the

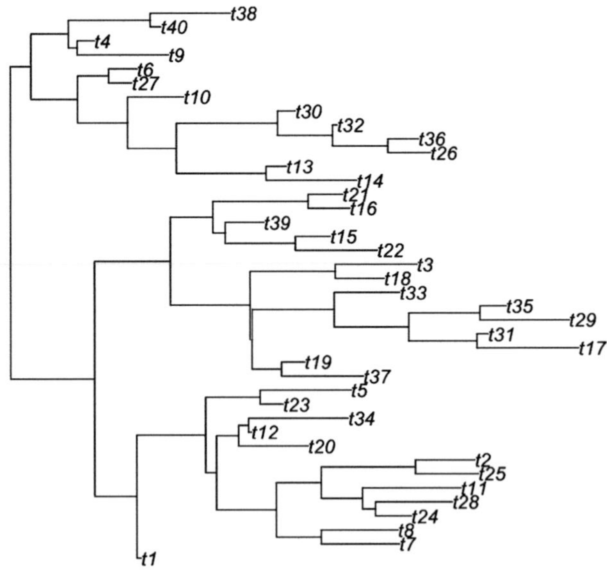

Fig. 2.10 A plot of a simulated phylogeny containing 40 species generated by randomly splitting lineages. Note that your phylogeny will be randomly generated and therefore will not look like the phylogeny plotted presently

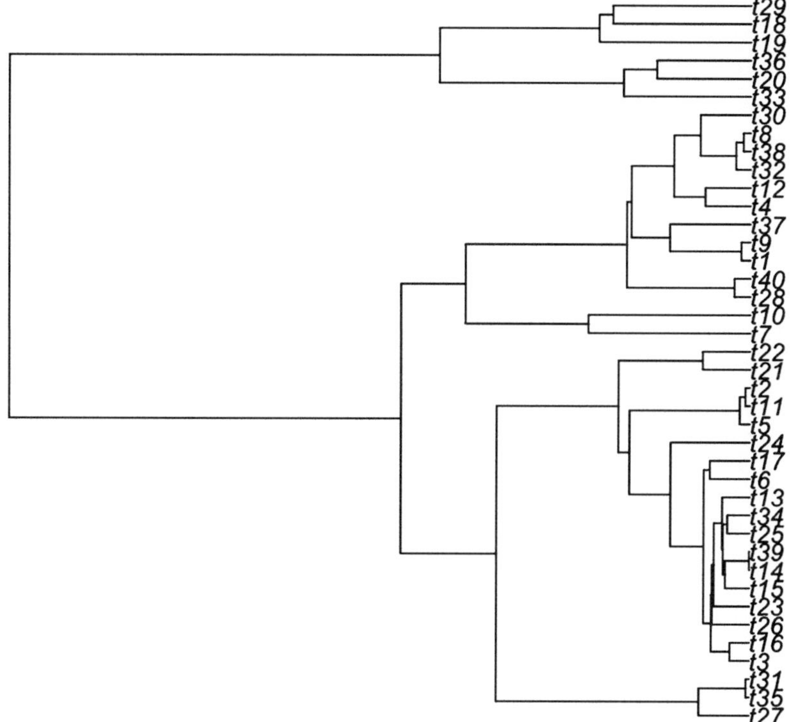

Fig. 2.11 A plot of a simulated phylogeny containing 40 species generated by simulating coalescence. Note that your phylogeny will be randomly generated and therefore will not look like the phylogeny plotted presently

terminal taxa are labeled t1, t2, ..., t40. This nomenclature for terminal taxa will be used in both simulation approaches described here. If you were simultaneously simulating community datasets with 40 taxa with a different naming convention, you could simply replace the names on the phylogeny with your list in the desired order using `new.tree$tip.label <- c(your.vector.of.names)`. If you wished to provide a vector of names prior to the simulation of the phylogeny to simply things, then you could provide that vector after the `tip.label` argument in the function.

The second basic approach for simulating phylogenies in R is to simulate a phylogeny that randomly clusters terminal taxa. Thus, in a sense this approach works backwards towards the tips while the previous method moves in the other direction. This "coalescence" method can be performed using the `rcoal()` function in the *ape* package.

```
> new.ultra.tree <- rcoal(40)
```

This generated a random coalescent tree. Now plot your new simulated coalescent tree (Fig. 2.11):

```
> plot(new.ultra.tree)
```

You will see a rooted and ultrametric phylogeny has been randomly generated. As I mentioned above, these are the two most basic approaches for simulating phylogenies, and if you were interested in performing a simulation study it would be useful to consider the assumptions you would want to make regarding the phylogeny prior to simulating anything and to decide whether these assumptions are met using one of the two approaches above or any other approaches available in R.

Both of the above simulation approaches generate one phylogenetic tree at a time. What if you wanted to generate 100 random phylogenies with 10 species each? You would not want to sit at your computer and enter the line of code 100 times. To perform any function repeatedly in R you can write a loop function called a `for()` loop. Such loops are generally fine for computationally easily problems like simulating a phylogeny with ten species, but for larger problems the `replicate()` function may be quicker. Though for simplicity we will just use a `for()` loop here and only produce nine random splitting phylogenies with ten terminal taxa in each and write each phylogeny to your working directory with a unique file number with a ".txt" file ending. If you are new to R you will see lines of code starting with a "#" symbol. This means that the text in that line following the symbol is a comment which R will disregard. Commenting your code is essential particularly when writing loops or functions. This helps you understand what you are doing and your goals and it also greatly helps anyone that may use your code in the future.

```
> for(i in 1:9){

    ## Make a random tree with 10 terminal taxa
    temp.random.tree <- rtree(10)

    ## Write the random tree to your working
    ## directory with each new file having a new
    ## number
    write.tree(temp.random.tree, paste(i,".txt",
    sep=""))

##close the loop
}
```

If you now navigate to your working directory on your hard drive you will see nine new files named "1.txt," "2.txt," ..., "9.txt." Open one of the files in a text editor and you will see a Newick version of your simulated random splitting phylogenetic tree for that iteration of the `for()` loop. Thus, it can be fairly easy and fast to simulate a phylogeny in R and again the central issue will be whether the simulation you are using is appropriate for your particular research goals. It is not possible for me to recommend one simulation approach over another for every problem because the goals of simulation approaches vary from study to study. Thus, my goal here is to simply introduce you to two basic simulation approaches to get you started and to demonstrate that simulating phylogenetic data is quite simple in R.

2.6   Conclusions

We have now covered the basics of phylogenetic information in R particularly focusing on the "phylo" class. There is a much more extensive text in the UseR! series on phylogenetic information in R that I recommend for more detail [15], but for the analyses in the present book this chapter serves as a sufficient primer. The most important aspects from the present chapter that you should keep at the forefront of your mind as we proceed into the next chapter on phylogenetic diversity have to do with how branch length information is stored and manipulated in the "phylo" class.

2.7   Exercises

1. Make a Newick file for five species (speciesA, speciesB, speciesC, speciesD, and speciesE) where speciesA and speciesE are most closely related to one another, speciesB and speciesD are most closely related to one another, and speciesC is most closely related to speciesA and speciesE. The file should have no branch lengths. Read this file into R and plot it with blue branches.
2. Take the Newick file you just made and put branch lengths in the file. Set all branch lengths to be 3.00 units in length and calculate the total branch length of the phylogeny (i.e., the tree length).
3. Calculate all subtrees for your phylogeny. Next choose one subtree and calculate the length of all the branches of that subtree and divide this number by the sum of the branch lengths in the whole phylogeny.
4. Use the `sample()` function to randomize the tip labels on a phylogeny where no name is lost and no name is used twice in the new phylogeny (i.e., sample without replacement).
5. Take the Newick file you made in Exercise 2 above and print a Nexus version of the file to your R console.
6. Repeat 1–5 above, except first generate a six species phylogeny speciesA–speciesF. speciesA and speciesB should be most closely related to one another, speciesC and speciesD should be most closely related to one another, and speciesE and speciesF should be most closely related to one another. Finally, speciesE and speciesF should be more closely related to speciesA and speciesB and more distantly related to speciesC and speciesD.
7. Simulate 45 coalescent trees (10 species in each) and write them to your working directory.
8. Put the 45 random coalescent tree files into a new directory, change your R working directory to that new directory and read each phylogeny into R, and calculate the total tree length for each automatically (i.e., do *not* do this one tree at a time—automate it). You will need to use the `list.files()` command and make a `for()` loop.

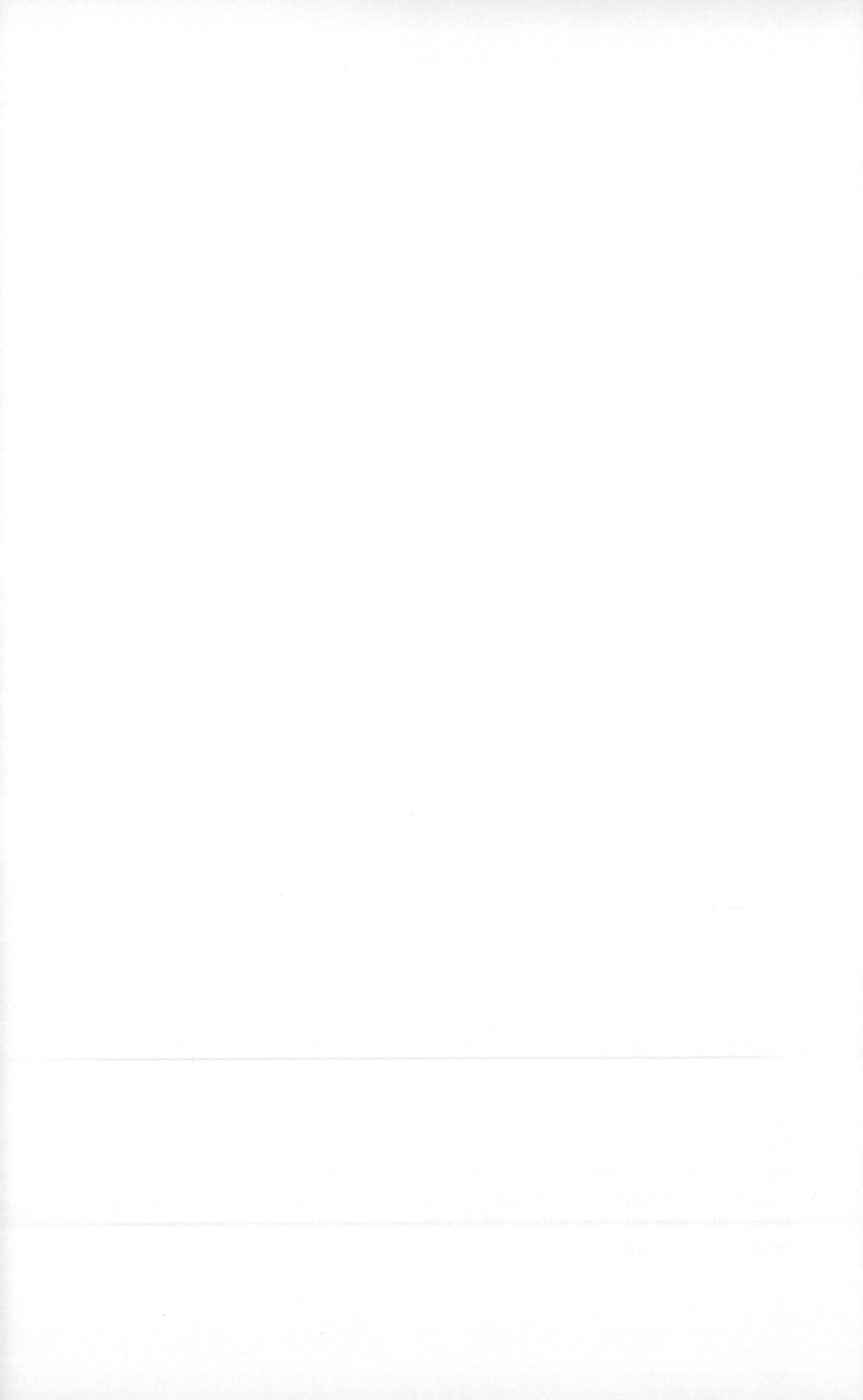

Chapter 3
Phylogenetic Diversity

3.1 Objectives

The objectives of this chapter are to first build a background understanding of why ecologists are interested in quantifying phylogenetic diversity (PD) and then to utilize R to quantify a variety of PD metrics that are the most frequently used. As in other chapters, we will focus on breaking down each analysis into its constituent parts to deepen our understanding of what exactly is being calculated and to facilitate your ability to write modifications of the code or novel code to generate PD analyses suited to your particular research objective.

3.2 Background

Ecologists interested in studying and conserving biodiversity are tasked with quantifying that diversity through space and time. Typically, this has been done using a measure of species diversity. Other dimensions of biodiversity such as phylogenetic diversity and functional diversity are less often quantified, but these forgotten dimensions may be equally or more important [1]. For example, conservation biologists are interested in phylogenetic measures of biodiversity as a way to provide a more robust estimate of the overall evolutionary history being currently preserved in protected lands or the potential amount of biodiversity that could be lost in threatened regions [25–27]. Basic community ecology research, on the other hand, has focused on the phylogenetic structure of communities in order to gain insights into their assembly (e.g., [28–38]). Despite their differing aims, both research programs generate estimates of the phylogenetic diversity (PD) within and between species assemblages across scales.

The online version of this chapter (doi: 10.1007/978-1-4614-9542-0_3) contains supplementary material, which is available to authorized users

N.G. Swenson, *Functional and Phylogenetic Ecology in R*, Use R!,

DOI 10.1007/978-1-4614-9542-0_3, © Springer Science+Business Media New York 2014

A large number of present day and past studies of community assembly and species coexistence focus on the relative importance of species similarity. In particular, empirical and theoretical research often asks whether ecologically similar species should tend to co-occur or not (e.g., [39, 40]) or whether the ecological similarity of species needs to be invoked at all to predict species coexistence and community assembly (e.g., [41–43]). For example, the mechanism of abiotic filtering proposes that ecologically similar species should co-occur [40] while strong biotic interactions (positive or negative) should lead to ecologically dissimilar species coexisting [39], although this viewpoint is challenged [44]. Lastly, a neutral model suggests that the ecological similarity of species need not be invoked to understand coexistence and assembly and that coexisting species are likely to be random with respect to their ecological similarities [41–43]. With these central predictions in place, the challenge becomes how can we measure the ecological similarity of species.

The phylogenetic diversity of communities has been of interest in community ecology for almost 100 years where early studies analyzed the ratio of species and genera in communities as a way to understand whether biotic or abiotic interactions are important in community assembly [45–48]. Specifically, a low species to genus ratio indicates the coexistence of distantly related species—what is termed today as phylogenetic overdispersion. A high species to genus ratio indicates coexistence of closely related species—what is termed today as phylogenetic underdispersion or clustering. This species to genus ratio approach continued for decades culminating with the famous community assembly rules and null model debates of the 1960s and 1970s (e.g., [49–52]). The foundation of the species to genus ratio approach is the assumption that closely related species are more likely to have similar niches— often termed phylogenetic niche conservatism (e.g., [53–55]). If closely related species tend to share similar niches, then community assembly via abiotic filtering should result in phylogenetic clustering whereas community assembly mediated by biotic interactions should result in phylogenetic overdispersion (e.g., [7, 30]). Charles Darwin originally alluded to niche conservatism when he considered the implications of common descent. Specifically, species that share a recent common ancestor should, on average, tend to be more similar to one another than they are to more distant relatives. If this assumption is supported then phylogenetic diversity may adequately estimate the functional diversity of an assemblage.

A problem with the species to genus ratio approach, beyond the assumption of niche conservatism, is taxonomic ranks do not convey detailed information regarding the time since two species diverged. A solution to this problem is to use phylogenetic trees with branch lengths. The branch lengths can be used to provide a more refined measure of relatedness between taxa and therefore PD. Though, until the early 2000s generating phylogenetic trees representing communities (i.e., community phylogenies) was considered not possible. Pioneering work by Cam Webb and colleagues that developed software tools such as Phylomatic [56] for estimating phylogenetic trees for plant communities largely removed this obstacle. This innovation sparked a large number of investigations into the phylogenetic relatedness of coexisting plant species primarily in the tropics where measurements of species function or niches in tens to hundreds of locally coexisting species are difficult to achieve at best [28, 57]. This work has primarily sought to quantify the phylogenetic diversity in a community and to ask whether that phylogenetic diversity is higher or lower than that randomly expected given their species diversity. These results are then often used to determine to what degree abiotic or biotic interactions govern community assembly, but one must be aware that the phylogenetic proxy for similarity is often not robust (e.g., [30, 34, 35, 44].

3.3 "Community" Datasets

A fundamental component of many of the analyses in this book are community data. I will use the term "community" throughout the book very loosely to refer to a group of species or taxa that you are studying. I will also interchange this term with "assemblage" or "site" or "sample." The interchangeability of these terms is not used here to reflect biological similarity. Rather, I use them interchangeably because the code can be used on a "community" of ten grass species, an "assemblage" of mammals on a continent, a group of insects found at a "site" on the side of a mountain, or the microbes found in a "sample" of soil. Ultimately, the point is that I do not wish readers to think the following analyses can only be utilized for "communities" however one wishes to define them or not.

Ecologists store their community or assemblage data in their brains, notebooks, and computers differently. Indeed, it often seems that the only commonality is a lack of similarity in how ecologists record and store their community datasets. The problem with this situation is that computer code written by others to analyze community data often requires a particular data format. This leaves ecologists the task of trying to determine how to wrangle their, sometimes large, datasets into the "right" format. The objective of this section is not to discuss how to wrangle ones community data. Rather I will outline the general community data format utilized in the R functions here and elsewhere and how to read your data into R so that it conforms to this format style.

The general format used to store and analyze community data in R is the "site-by-species" matrix. I generally will refer to this as the community data matrix. The "site" can be a community, an assemblage, a sample, or whatever group of taxa you wish to analyze. The sites are arranged in rows with the names of the rows being the names of the sites. Thus, if you had sampled insects every 100 m up a mountain for a total of ten sites, your community data matrix would have ten rows. The columns of your community data matrix contain species or taxa found in your sites. For example, if you found 94 insects total in your 10 sites on the mountainside, you would have a community data matrix with 10 rows and 94 columns with the row names being the unique names of sites and the column names as the unique names of species or taxa. If your data were collected on the appropriate scale, you may then consider that your matrix contains many communities (i.e., individual rows) within a meta-community (i.e., the entire matrix). The values in the matrix could be binary indicating the presence or absence of each taxon in each site or the values could be continuous indicating the abundance or relative abundance of each species or taxon in each site. We can examine these data types by reading in examples of each. First we read in a community data matrix where the values are binary indicating the presence or absence of each species. The example table is tab delimited with column headers and the row names in column one.

```
> pa.matrix <- read.table("pa.matrix.txt", sep =
"\t", header = T, row.names = 1)
```

Summing the rows of this matrix will provide the richness for each site.

```
> rowSums(pa.matrix)
```

Summing the columns of this matrix would provide the number of sites a taxon occupies. This value could be divided by the number of sites (i.e., the number of rows) to calculate the occupancy rate for a taxon in the study system.

```
> colSums(pa.matrix)
```

If we examine a community data matrix with values indicating the number of individuals, we can see that the row sums are the total abundance of all taxa per site and the column sums are the total abundances of taxa in the study system.

```
> abund.matrix <- read.table("abund.matrix.txt", sep =
"\t", header = T, row.names = 1)

> rowSums(abund.matrix)

> colSums(abund.matrix)
```

If we examine a community data matrix with values indicating the relative abundance of taxa, we can see that the row sums should equal one and that the column sums equal the sum of the relative abundances for a taxon across sites.

```
> ra.matrix <- read.table("ra.matrix.txt", sep = "\t",
header = T, row.names = 1)

> rowSums(ra.matrix)

> colSums(ra.matrix)
```

The community data matrix format is not unique to phylogenetic and functional analyses of ecological data. Rather, phylogenetic and functional analyses have adopted this format in part because the format allows for many aspects of communities to be quantified quickly and because maintaining this format allows phylogenetic or functional analyses to utilize R functions originally built for other ecological analyses.

Some ecologists may record their data in the field naturally as a site-by-species matrix and enter it into their computers in that format. In such instances it is easy to read the data into R for ecological analyses. In many other cases, ecologists record their data giving each individual its own row in a field notebook with perhaps a column indicating the individuals name, the location of the individual (i.e., what community it was found in), the size of the individual, and so on. This type of data can be read into R relatively easily and automatically recompiled as a community data matrix. For example, imagine you could easily format your data with three columns where each row is an individual. The first column is the name of the site where the individual was observed, the second column was a one for present or perhaps the size of the individual, and the third column was the name of the species or taxon for the individual. Such a format is likely easy to generate by moving columns in the typical spreadsheet programs many ecologists use and to save as a tab delimited .txt file with no column names. This data format can be read into R using the *picante* package [58]. First load the *picante* package.

```
> load(picante)
```

Next use the `readsample()` function to read in your three-column text file with no column headers and examine the output.

```
> my.3.sample <- readsample("matrix.3col.txt")
```

In this instance I put ones in the second column to simply indicate presences of an individual. The values in the community data matrix are the numbers of individuals for a taxon per site. If I utilized the size of the individual in the second column, the values would have been equal to the sum of the sizes of all individuals for a taxon per site. If we wished to simplify our community data matrix such that the values are binary representing presence or absence, we could use the *vegan* package `decostand()` function and use the "pa" method.

```
> decostand(my.3.sample, method = "pa", MARGIN = 1)
```

Similarly, if we wanted to transform our original matrix into relative abundance we can use the same function and the "total" method, which divides each value in a row by the total of all values in a row if the MARGIN is set to one.

```
> my.ra.3.sample <- decostand(my.3.sample, method =
"total", MARGIN = 1)

> my.ra.3.sample
```

If you would like to save a copy of your community data matrix, you can write it to your hard drive using the `write.table()` function setting the value separation to "\t" for tab delimited text and row and column names as True.

```
> write.table(my.ra.3.sample, "my.ra.3.matrix.txt",
sep = "\t", row.names = T, col.names = T, quote = F)
```

If you would like to save your community data matrix in the three column format, you can use the `writesample()` function.

```
> writesample(my.ra.sample, "my.new.3col.data.txt")
```

Now that we understand how community data can be read into R and stored in a matrix format we can move on to the analysis of our community data. In the following sections we will focus on calculating three main classes of PD metrics using community data matrices. The goal is to be able to simultaneously calculate the PD of each site (i.e., row in the community data matrix) by providing the R function the community data matrix and a phylogenetic tree.

3.4 Tree-Based Measures of Phylogenetic Diversity

Metrics of biodiversity that incorporated evolutionary history have existed for approximately a century. Early metrics utilized taxonomic ranks and taxonomic ratios as phylogenetic trees with reasonably inferred branch lengths were not available [45–48]. Over the past 30 years the development of phylogenetic diversity metrics that utilize branch length information has rapidly accelerated and the number of articles published in leading ecological journals that utilize taxonomic ratios is nearly zero. Dan Faith's "PD" metric published in 1992 [25] is often considered the first clear demonstration of how to calculate phylogenetic diversity using a phylogenetic tree with branch lengths, though previous discussions do exist (e.g., [59]). Faith's metric was, and still often is, called "PD." This name causes problems given that researchers often abbreviate the general concept or measurement of phylogenetic diversity, and not Faith's metric per se, as "PD." I prefer to utilize the abbreviation "PD" to refer to the general concept or measurement and "Faith's Index" as the metric published by Faith [25] and will do so throughout this book. Faith's Index is described as the sum of the branch lengths connecting all species in an assemblage.

$$Faith = \sum_{i}^{n} l_i$$

where there are n branches with each having a length of l in a phylogeny containing only the species in the assemblage. Thus, adding a species to an assemblage adds at minimum a terminal branch to the phylogeny thereby often leading to a correlation between the species richness of an assemblage and Faith's Index in a study system. Given that Faith's Index is the sum of all branch lengths in a phylogeny containing only the species in an assemblage, the calculation of this metric is simple if we can quickly prune a phylogenetic tree to only include the species in our assemblage. We begin by reading our community data matrix into R.

```
> my.sample <- read.table("PD.example.sample.txt", sep
= "\t", row.names = 1, header = T)
```

In this example file the first row contained the taxa names and the first column contained the name of the assemblage. The example phylogenetic tree containing all of the species in our community data can be read into R using the `read.tree()` function.

```
> my.phylo <- read.tree("PD.example.phylo.txt")
> my.phylo
```

Fig. 3.1 A plot of our
example phylogeny

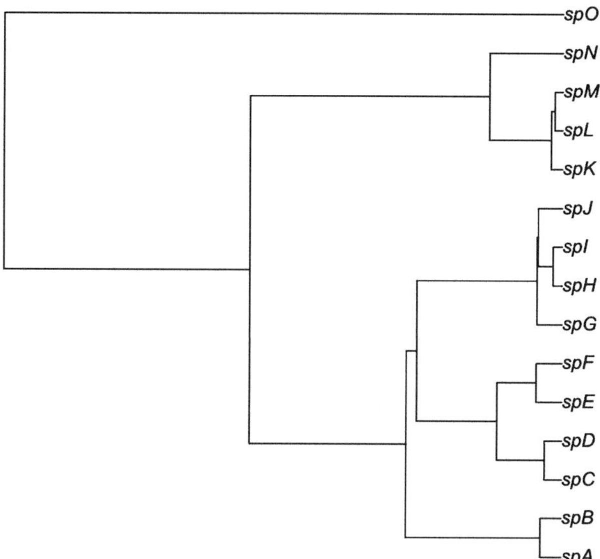

We see that the phylogeny contains 15 taxa (i.e., tips) and 14 nodes, it has branch
lengths, and it is rooted. Because the number of nodes is one less than the number
of tips, we know the phylogeny is fully bifurcating. Generally, when reading any
data into R you should plot it and phylogenies are no different. This helps identify
any clear problems with the dataset prior to any additional analyses that may be
affected (Fig. 3.1).

```
> plot(my.phylo)
```

The phylogeny has now been plotted and we do not see any clear problems with the
dataset. Potential problems that could easily be detected are nonsensical taxa names
on the tips of the tree that do not match the names of taxa in your community data
matrix; the phylogeny is not ultrametric when you expected it to be ultrametric (this
can also be checked using `is.ultrametric()`, etc.

In this simple example for calculating Faith's Index for a single community, and
not all communities at once, we must extract the data from the community for all
species that are present (i.e., have an abundance or relative abundance greater than
zero). The community we will analyze is the first community in our example dataset. We will therefore ask R for the first row in our example community data matrix
for only those columns (i.e., species) where the values (i.e., the abundance or relative abundance) are greater than zero.

```
> my.sample[1, my.sample[1, ] > 0]
```

The ones in this code refer to the first row in the data matrix and the greater than zero
refers to wanting only those columns with a value greater than zero. The result is a

vector containing the abundances or relative abundances of only the species present in the first community with the names of the vector being the species names. Because only the species present in the community are in this output, we can simply calculate the species richness of the community by asking R for the length of this vector.

```
> species.richness.com.1 =
length(my.sample[1, my.sample[1, ] > 0])

> species.richness.com.1
```

Now that we know how to extract from our community data only those taxa present in a community, we can utilize this information to prune our phylogeny that contains all of the taxa present in our community and many taxa that are not present. This can be accomplished by using the `treedata()` function in the R package *geiger*. To do this we first load the *geiger* package [60].

```
> library(geiger)
```

We now use the `treedata()` function to prune our original phylogeny. This function is designed to match and organize phylogenetic and trait data for comparative analyses. Specifically, the function requires a phylogenetic tree and a matrix or vector of trait data with the row names for a matrix or names for a vector being the taxa names. The function returns pruned versions of the phylogeny and trait data. Although we are not using actual trait data presently, we can utilize the vector of abundances and names of taxa extracted from our community data matrix as the "trait" data thereby pruning our large original phylogeny to only contain the species present in our first community.

```
> treedata(my.phylo, names(my.sample[1, my.sample[1, ]> 0]))
```

The output includes a pruned phylogeny and trait dataset as well as a warning message. The warning message in this instance is telling us that several taxa (i.e., tips) in the phylogeny were not found in the trait dataset and were therefore pruned from the phylogeny. It then lists these taxa. Given that the "trait" data we provided the function was the abundance of only the present species in our community, this warning actually confirms we have accomplished our goal of pruning taxa from the phylogeny not in our community. Because the warnings will appear for every community in our dataset and we are only interested in obtaining the pruned phylogeny for a community, we can modify the above code as follows.

```
> pruned.tree <- treedata(my.phylo, names(my.sample[1,
my.sample[1, ] > 0]), warnings = F)$phy

> pruned.tree
```

We now have obtained a pruned version of our larger phylogeny. This pruned version only contains the species found in our community. To compare this pruned

Fig. 3.2 A plot of our example phylogeny after being pruned to only include species found in our community

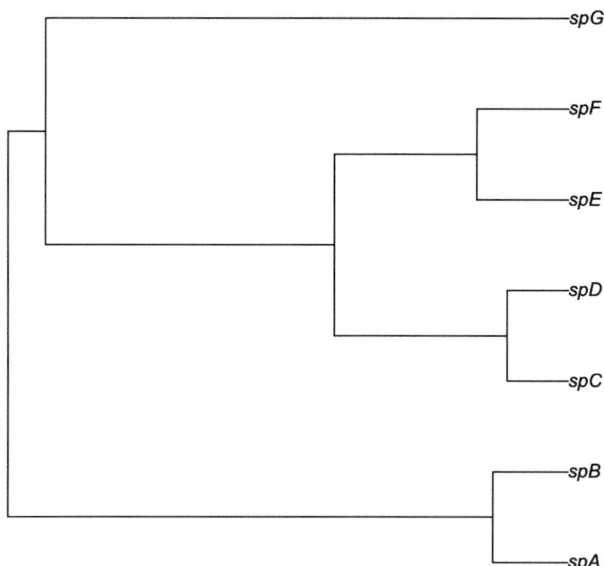

phylogeny to the original one, open a new quartz window and plot the pruned phylogeny (Fig. 3.2).

```
> quartz()
> plot(pruned.tree)
```

Faith's Index can now be easily computed using the pruned phylogenetic tree containing only species found in the community by summing all branch (i.e., edge) lengths.

```
> sum(pruned.tree$edge.length)
```

This code only provides the Faith's Index for a single community, but we would ideally like to calculate the Faith's Index for all communities at once. To accomplish this we first write a simple function named `prune.sum.function()` for pruning a large phylogenetic tree called `my.phylo` using row x from the community data matrix. We then sum the branch lengths in the pruned phylogeny and output the result.

```
> prune.sum.function <- function(x){

        tmp.tree <- treedata(my.phylo, x[x > 0])$phy
        sum(tmp.tree$edge.length)

    }
```

Given that `prune.sum.function()` takes a single row (i.e., community) from the community data matrix as input, we can apply this function to all rows simultaneously using the `apply()` function using a MARGIN value equal to 1 to indicate that we want to apply the `prune.sum.function()` to each individual row in the input matrix called `my.sample`.

```
> apply(my.sample, MARGIN = 1, prune.sum.function)
```

We have now rapidly calculated the Faith's Index value for each of our communities. As is the case for many of the analyses that we will conduct in this book, a function already exists to calculate the Faith's Index metric. The phylogenetic and function diversity functions are primarily available in the *picante* package. However, a difficulty with the functions written in this package is that they often rely on `for()` loops, which can be considerably slower than `apply()` functions. For example, it is common for functions in *picante* to loop through each row in a community data matrix rather than simply use an `apply()` function as we have done above. It is known that `for()` loops slow down analyses, and `apply()` type functions should be used as much as possible. While the difference in computing speed for small datasets is generally negligible, the difference in speed when handling large datasets can be considerable particularly when randomization procedures are invoked. In such instances I recommend not using the *picante* package and simply using the `apply()` based functions provided above and elsewhere. Nonetheless, calculating Faith's Index using *picante* can be done using the `pd()` function.

```
> pd(my.sample, my.phylo, include.root = F)
```

The original version of Faith's Index did not include the root of the pruned phylogenetic tree for a community or assemblage, but recently Faith's Index has been stated to include the root. There is even a differently named metric called "Evolutionary History" or "Evolutionary Heritage" (EH) that was designed to be different from Faith's Index by its inclusion of the root [61]. The rationale for including the root is that it provides more information regarding the long evolutionary history leading up to the species found in the community. Thus, those interested in calculating the historical diversity in an assemblage for conservation purposes, for example, the inclusion of the root may be preferred. This can easily be calculated as follows.

```
> pd(my.sample, my.phylo, include.root = T)
```

A critique of Faith's Index has been that it does not include information regarding the relative abundance of individual species in the assemblages being analyzed. This lack of information may not be of as much interest for conservation assessments, but it may be terribly important for analyses of community structure and diversity. This has led to the development of a version of Faith's index that is weighted by abundance [62] that I will call the Weighted Faith's Index.

$$Weighted\,Faith = n \times \frac{\sum_{i}^{n} l_i \bar{A}_i}{\sum_{i}^{n} \bar{A}_i}$$

where n is the number of branches in the phylogenetic tree, l_i is the length of the ith branch, and $\overline{A}_i$ is the average abundance of all species subtended by that branch. As you can see, calculating this weighted metric is more complex than simply summing the branch lengths in a phylogeny containing community members. It requires calculating a metric over all individual branches in the phylogeny containing the community members. To calculate the Weighted Faith's Index for a single community, we start by extracting from a larger phylogeny the phylogeny containing only the species in our community. In this example we will analyze the first community (i.e., species with an abundance greater than zero in row one of our `my.sample` object).

```
> com.1.phylo <- treedata(my.phylo, names(my.sample[1,
my.sample[1, ] > 0]))$phy
```

The next step is to generate an empty matrix that will hold the output for the variables we would like to quantify for each individual branch in our community phylogeny. We will fill the matrix initially with NA values, have four columns and the same number of rows as there are edges (i.e., branches) in our community phylogeny. The number of rows therefore corresponds to the parameter n in the above equation.

```
> branches <- matrix(NA, nrow = nrow(com.1.phylo$edge),
ncol = 4)
```

In the first two columns we will place the beginning and ending (i.e., basal and terminal) node for each branch in our community phylogeny by asking for the edges stored in our community phylogeny object. We will use this information to ask R what species are subtended by each branch on the phylogeny.

```
> branches[,1:2] <- com.1.phylo$edge
```

In the third column of the matrix we will place the length of all edges (i.e., branches) in the phylogeny. This column therefore contains the parameter l_i in the above equation.

```
> branches[,3] <- com.1.phylo$edge.length
```

```
> head(branches)
```

We now see that we have the nodes defining each edge in the phylogeny and the length of that edge. To visualize what the node numbers in the first two columns correspond to in our phylogeny, we can first plot the phylogeny and then plot the internal node labels followed by the terminal (i.e., tip) node labels (Fig. 3.3).

```
> plot.phylo(com.1.phylo, show.tip.label=F)
```

```
> nodelabels(bg = "white")
```

```
> tiplabels(bg = "white")
```

Fig. 3.3 A plot of our example phylogeny with the internal and terminal nodes labeled. The node numbers displayed help to understand how we can select individual branches for the Weighted Faith's Index calculation

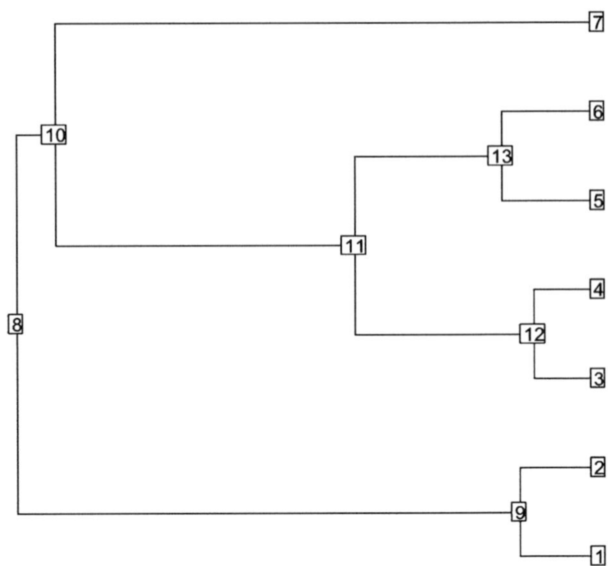

We can now visualize what the information in the matrix we have created indicates. We can see that an internal node in our phylogeny splits into two branches, if the tree is completely bifurcating, each of which ends in a different node. For example, in this phylogeny node 17 branches and ends in the terminal nodes 27 and 18. Thus, if we wanted only the species from one of those two branches, we would not ask for all species subtended by node 17. Rather we would ask for all species subtended by either node 27 or node 18. In other words we would like to ask R to report the average abundance of all species descendent from node 27 or 18 and not node 17 (i.e., the terminal node numbers in column two of our `branches` matrix object). We could therefore write a simple `for()` loop to go through each row in our matrix, ask for the species descendent from the node in column two, and average their abundance. Summing this information would provide the denominator in the above equation.

```
> for(i in 1:nrow(branches)){

        ## Using the terminal node number for each branch
        ## in column two of the node.bls object we pull
        ## out the taxa names subtended by the branch
        ## described in each row of the node.bls object.
        leaves.node <- tips(com.1.phylo, branches[i,2])

        ## Average the abundance of the species found in
        ## community 1, which is the first row in our
        ## my.sample object. This is equal to Ai in the
        ## raw weighted.Faith equation.
        branches[i,4] <- mean(my.sample[1,leaves.node])

    ## close loop
    }
```

As noted above, the number of branches in the phylogeny is the *n* parameter in the equation and is simply the number of rows in our matrix or the number of rows produced when asking for the edges from a phylo object.

```
> number.of.branches <- nrow(com.1.phylo$edge)
```

The denominator of the Weighted Faith's Index is calculated by summing column four in our matrix. Recall that each row in column four contains the average abundance of the species subtended by the node number indicated in column two. In other words, each row in column four is an A_i value.

```
> denominator <- sum(branches[,4])
```

The numerator for the Weighted Faith's Index equation is the sum of the products of individual branch lengths (l_i) found in column three of our matrix and the average abundance of the species from that branch (A_i) found in column four.

```
> numerator <- sum(branches[,3] * branches[,4])
```

We now have the values to calculate the Weighted Faith's Index for a single community.

```
> weighted.faith.output <- number.of.branches *
(numerator / denominator)
```

While useful, the above approach only analyzes a single assemblage at a time. This is obviously suboptimal and was only used to demonstrate the general approach. A better method would be to write a function to calculate the Weighted Faith's Index for a community and to apply this approach to the rows in the community data matrix. To do this we write two small functions. The first is to generate a function called `get.leaves()` that will provide the species subtended by each branch. This will be accomplished by providing a temporary community phylogeny and the values in the second column of the `branches` object.

```
> get.leaves <- function(x){

    leaves.node <- tips(tmp.tree, x[, 2])

}
```

The second function which we will call the `weighted.faith.function()` will take an input community data matrix and trim a larger phylogeny called `my.phylo` to an individual community phylogeny for each row in the community data matrix. Using the community phylogenies and the `get.leaves()` function, we can calculate the Weighted Faith's Index.

```
> weighted.faith.function <- function(x){

    ## extract the names of species in a community
    ## with an abundance greater than zero and use
    ## that information to make a pruned phylogeny
    ## for that community.
    tmp.tree = treedata(my.phylo, x[x > 0])$phy

    ## Create empty branches matrix
    branches <- matrix(NA, nrow = nrow(tmp.tree$edge),
    ncol = 4)

    ## Fill first two columns of the matrix with node
    ## numbers defining each edge.
    branches[,1:2] <- tmp.tree$edge

    ## Fill the third column with the length of each
    ## branch
    branches[,3] <- tmp.tree$edge.length

    ## Apply the get.leaves() function to each row in
    ## the branches object. This will retrieve
    ## species names subtended by each branch (i.e.
    ## row) in the branches matrix
    leaves <- apply(branches, MARGIN = 1, get.leaves)

    ## Now quickly loop through each set of leaves to
    ## calculate the mean abundance (Ai) for those
    ## species.
    for(i in 1:length(leaves){
        branches[i,4] <- mean(x[leaves[i]], na.rm = T)

    }

    ## Lastly calculated the Weighted Faith's Index
    nrow(tmp.tree$edge) * ((sum(branches[,3] *
    branches[,4])) / (sum(branches[,4])))

}
```

While still a little cumbersome and containing a loop, the above code can now be quickly applied to the community data matrix, `my.sample`, using the `apply()` function.

```
> all.weighted.output <- apply(my.sample, MARGIN = 1,
weighted.faith.function)

> all.weighted.output
```

We have now completed the calculation of the Weighted Faith's Index. To my knowledge this function is not available in any existing R package. Although there are likely to be many tree-based metrics such as Faith's Index developed in the future, the original metric and its abundance-weighted derivative are likely to remain the most widely used in conservation. Since the original derivation of Faith's Index in 1992, more complex measures of phylogenetic diversity have been developed and that majority of these utilize some variety of phylogenetic distance matrix. In the next section we will discuss and implement the two main classes of distance-based measures of phylogenetic diversity.

3.5   Distance-Based Measures of Phylogenetic Diversity

The success and widespread use of tree-based measures of phylogenetic diversity (PD) lies in their precedent and the ease with which summing branch lengths or weighted branch lengths can be understood. Despite these qualities several additional metrics have been published that can produce alternative, or maybe even refined, insights into the phylogenetic structure of assemblages. The majority of these new metrics rely on a phylogenetic distance matrix of some variety and generally arose from a tradition in trait-based ecology of examining and contrasting the pairwise functional distance between species or individuals in a community or the nearest functional neighbor distances. Thus, in many ways the same mathematics used to calculate early functional diversity metrics were simply adopted to quantify phylogenetic diversity replacing a trait distance matrix with a phylogenetic distance matrix. Although it seems like a handful of new phylogenetic diversity metrics are published every year at this point, most generally fall into one of two categories— pairwise or nearest neighbor. Some "new" metrics may not exclusively use these conceptual frameworks, but they are often highly correlated with existing metrics in one of these two categories and it can often be difficult to discern what novel information is being extracted by the "new" metrics published. Given this situation, I will simply cover the two main classes of PD metrics and the classic metrics representing these classes. In some cases I will point you to associated metrics that I will not discuss in as much detail but can be easily calculated in R.

3.5.1   Pairwise Measures

Distance-based measures of phylogenetic diversity utilize a phylogenetic distance matrix or phylogenetic variance–covariance matrix to quantify a metric of relatedness between species or taxa in a community or sample. Phylogenetic distance matrices are simply matrices with taxa names on the rows and columns and values in the cells depicting the phylogenetic branch length separating each pair of species. Thus, the diagonal in such a matrix is the distance from a taxon to itself and therefore zero. A phylogenetic distance matrix can be calculated in R using a phylogeny object as class "phylo" and the `cophenetic()` function (Fig. 3.4).

```
> plot.phylo(my.phylo)
```

```
> dist.mat <- cophenetic(my.phylo)
```

```
> dist.mat[1:4, 1:4]
```

The phylogenetic distance matrix therefore simply reflects the branch lengths between taxa with no other underlying assumptions beyond those that have gone into original inference of the phylogenetic tree. A phylogenetic variance–covariance

Fig. 3.4 A plot of our example phylogeny

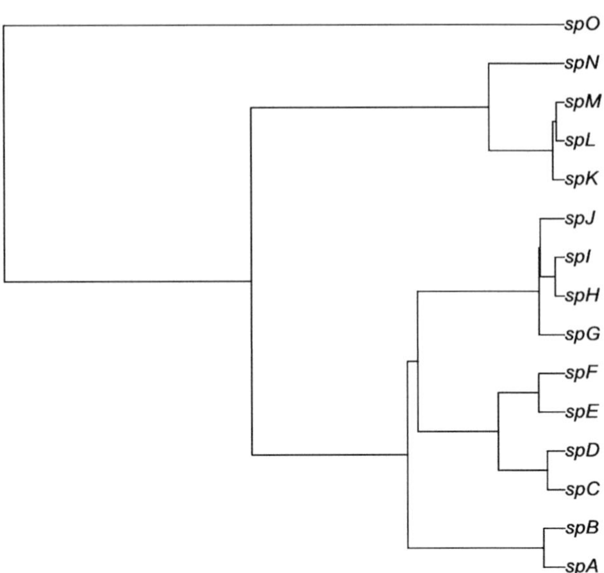

(VCV) is slightly different from a phylogenetic distance matrix. The phylogenetic VCV matrix represents the expected variance and covariances between species assuming a particular model of trait evolution. In the simplest case, which is the case generally utilized in phylogenetic diversity metrics relying on phylogenetic VCV matrices, is that traits are assumed to evolve under a Brownian Motion model. A Brownian Motion model assumes that traits evolve as a random walk along the length of a branch in a phylogeny. Thus, the potential variance in a trait increases with branch length. Further, the expected similarity between the traits evolving on two lineages increases with their total shared branch length such that the expected covariation of the trait values for two lineages increases with the shared branch length. For example, in a phylogeny with a root 100 million years old where species A and B diverged 12 million years ago and the lineage containing A and B diverged from the lineage containing species C 40 million years ago, the expected covariation of the trait values for A and B is proportional to 88 and the expected covariation of trait values for A and C or B and C is proportional to 60. These expected covariances make up the off diagonal values in the phylogenetic VCV matrix. The expected variance in a trait is proportional to the branch length from the root to the tip. In this case all species would have an expected variance proportional to 100. These expected variances make up the diagonal values in the phylogenetic VCV matrix. Although it is generally not the case for phylogenetic diversity analyses, other models of trait evolution may be used to generate the phylogenetic VCV matrix and that will change the formula for the expected variances and covariances.

A phylogenetic VCV matrix assuming Brownian Motion can be generated in R using a phylo object and the `vcv()` function in the *ape* package.

```
> vcv(my.phylo)
```

We see that the diagonal values are equivalent to the root-to-tip distance, and the off diagonal values indicate the shared branch lengths between all species in the phylogeny. Thus, when assuming a Brownian Motion model, we see that the phylogenetic VCV matrix is not all that different from a phylogenetic distance matrix for an ultrametric phylogeny. The exceptions being that the diagonals are root-to-tip distances and not zero, and the expected covariances are equivalent to the root-to-tip distance minus the one-half of the phylogenetic distance between two species.

Now that we have a basic understanding of the types of matrices that enter into distance-based PD metrics, we can proceed to discuss and calculate in R the most widely used metrics. We will begin with pairwise metrics of PD. Pairwise metrics have their roots in trait-based analyses where the idea is to summarize the overall average distance in uni- or multivariate trait space between all species combinations in a community or sample. This concept was adopted by Webb [28] and modified to simply use a phylogenetic distance matrix instead of a trait distance matrix. The pairwise metric that Webb [28] developed is still the most widely used metric and is called the mean pairwise distance (mpd).

$$mpd = \frac{\sum_{i}^{n}\sum_{j}^{n}\delta_{i,j}}{n}, \quad where\ i \neq j$$

where there are n species in the community or sample, δ is the phylogenetic distance matrix, and $\delta_{i,j}$ is the phylogenetic distance between species i and j. Because this metric calculates all pairwise distances in a community or sample, it is often considered to be a "basal" metric of PD. That is, this metric and all other pairwise metrics capture the overall phylogenetic dissimilarity of the taxa in a sample and do not detect finer scale phylogenetic patterns. We will discuss this in more detail in the next subsection on nearest neighbor metrics and why calculating both types of metrics and contrasting the results is useful.

The objective when calculating the mpd metric is to generate a phylogenetic distance matrix between all species in a community and to take an average not including the diagonal values because species i cannot equal species j. However, in most cases the phylogeny we are utilizing contains many additional species or taxa that are not found in our sample or community. Thus, we must prune this larger phylogeny to only contain the species in our sample or community. Here, we will first do this using a single community in our community data matrix. We will focus on the first community in our data matrix (i.e., row one). The first objective is to extract only those species in our community in row one that have a positive value indicating they are present. We will ignore their abundances since we are calculating the original version of mpd that weights all present species equally. To extract the present species in community one, we ask for all species in row one that have a positive value.

```
> com.1 <- my.sample[1, my.sample[1, ] > 0]

> com.1

> names(com.1)
```

We now have a vector of values containing only the names of the species present in our community. We also have the names of each species in this vector. We can use these names to extract only the rows and columns from the phylogenetic distance matrix we generated above. This will provide a phylogenetic distance matrix that does not contain the species not present in our community.

```
> dist.mat.com.1 <- dist.mat[names(com.1),
names(com.1)]
```

We now have a phylogenetic distance matrix for our community. We could simply take a mean of the upper or lower triangle of this matrix if we allowed species i to equal species j in our calculation, but since this is not allowed in the calculation of mpd we will use the as.dist() function to provide only the lower triangle without diagonal values and take a mean.

```
> mean(as.dist(dist.mat.com.1))
```

This simple calculation has now given us the mpd for a single community, but we would now like to calculate it simultaneously for all communities in our community data matrix. This can be accomplished by first generating a small function that takes a row from the community data matrix and does the calculation we just performed.

```
> new.mpd.function <- function(x){

      ## Get the names of species present in the
      ## community
      com.names <- names(x[x > 0])

      ## Calculate mpd by extracting the lower triangle
      ## of a phylogenetic distance matrix comprised of
      ## only the species in our community.
      mean(as.dist(dist.mat[com.names, com.names]))

  }
```

Now that we have a simple function to calculate mpd for a single row of the community data matrix, we can use the apply() function to apply this function to all rows in the community data matrix to rapidly calculate mpd for all communities.

```
> apply(my.sample, MARGIN = 1, new.mpd.function)
```

The mpd of all communities can also be calculated using a preexisting function in the R using the mpd() function in the *picante* package. The mpd() function is conceptually similar to the above aside from using a slower for() loop to calculate the metric across all rows in the community data matrix.

```
> mpd(my.sample, cophenetic(my.phylo),
abundance.weighted = F)
```

Although the vast majority of the early papers that employed the mpd metric
weighted all species or taxa in a community equally in the calculation, (e.g., [28, 32,
33, 63]), we can see from the code above that the mpd() function has an option to
weight the calculation by abundance. Weighting by abundance is often useful in the
analyses of communities since species are rarely equally abundant and the variation
in the shape of the abundance distribution across communities relays important ecological information. Further, weighting metrics such as mpd by abundance adds
another valuable piece of information particularly related to the phylogenetic distribution of abundance itself (see [38, 64–67]). For example, the most abundant species in a community may be very distantly or very closely related. In such instances
the abundance weighted mpd value would be much higher or much lower, respectively, than the unweighted mpd value.

The abundance weighted mpd, which I will call mpd.a, can be formalized as
follows:

$$mpd.a = \frac{\sum_i^n \sum_j^n \delta_{i,j} f_i f_j}{\sum_i^n \sum_j^n f_i f_j}, \quad where\ i \neq j$$

The variables in the mpd.a equation are the same as those in the mpd equation with
the addition of the abundances of species denoted with the variable f for frequency.
The mpd.a metric was "invented" and became more widely used towards the end of
the first decade of this millennium. However, it should be pointed out that Rao had
published a very similar metric nearly three decades earlier [68]. The Rao metric
was designed as a general dissimilarity metric utilizing a distance matrix between
taxa and has been applied to both phylogenetic and trait-based investigations. The
primary difference between the Rao metric for within community diversity and the
mpd.a calculation is that species i can equal species j. In simple terms, this means
that a mean of the lower triangle of a community phylogenetic distance matrix is
calculated using the diagonal elements.

$$Rao's\,D_{alpha} = \frac{\sum_i^n \sum_j^n \delta_{i,j} f_i f_j}{\sum_i^n \sum_j^n f_i f_j}$$

While the inclusion of zeros in the Rao calculation, indicating the distance from an
individual species to itself, is conceptually important we will see below that the
mpd.a and Rao's D_{alpha} metrics are monotonic and therefore will give the same rank-
ings of PD when comparing communities in your community data matrix. In other
words, conceptually it might matter, but it will not matter for your results.

In the spirit of consistency we will proceed with a simple calculation of mpd.a in R while understanding that this concept, and essentially the same metric, was invented by Rao [68]. The calculation of mpd.a is slightly more difficult because we must weight the mean of the phylogenetic distances separating all species in a community by the product of their abundances. We will eschew the calculation of the metric for a single community in this instance and simply write a function that can be applied to all rows in our community data matrix.

```
> new.mpd.a.function <- function(x){

    ## Get the names of the species in the community.
    com.names <- names(x[x > 0])

    ## Make a matrix with one row containing
    ## abundances and names of all present species
    com <- t(as.matrix(x[x > 0]))

    ## Make phylogenetic distance matrix for taxa in
    ## community.
    com.dist <- dist.mat[com.names, com.names]

    ## Calculate the product of the abundances of all
    ## species in the community.
    abundance.products <- t(as.matrix(com[1, com[1, ]
    > 0, drop = F]))%*% as.matrix(com[1, com[1, ] > 0,
    drop = F])

    ## Calculate a mean of the community phylogenetic
    ## distance matrix weighted by the products of
    ## all pairwise abundances.
    weighted.mean(com.dist, abundance.products)

}
```

The above function for the abundance weighted metric, mpd.a, can now be applied to all rows in the community data matrix to rapidly calculate the value for each community simultaneously.

```
> apply(my.sample, MARGIN = 1, new.mpd.a.function)
```

The mpd() function in the *picante* package can be used for this same purpose by switching the abundance.weighted argument to true. Again the general workings of the calculations are similar aside from the above code using an apply() function rather than a for() loop.

```
> mpd(my.sample, cophenetic(my.phylo),
abundance.weighted = F)
```

In addition to the metrics detailed above, several other pairwise metrics have been proposed in the literature. The vast majority produce results that are perfectly or monotonically related to the mpd-based measures above as we will see in Sect. 3.6. Here I will briefly discuss a few of these measures that are well known and often implemented in the literature. The first metric I will cover was produced by Hardy and Senterre [69] where they translated commonly used metrics from population genetics, such as F_{st}, into the realm of community phylogenetics. Specifically, Hardy and Senterre proposed that phylogenetic diversity, D, within a community k can be calculated as:

$$D_k = \sum_i^n \sum_j^n \delta_{i,j} \, f_{ik} f_{jk}$$

where the variables are the same as the above equations aside from using a subscript of ik to specify the abundance of species i in community k and not another community. This clause becomes important when Hardy and Senterre [69] extend their mathematical treatment to the partitioning of gamma phylogenetic diversity into the within and between community components. We can see that this metric should be highly correlated with both the Rao and mpd.a metrics and is a natural phylogenetic extension of the Shannon Diversity index [69]. We can also see that this metric could be simply calculated using the above code for mpd.a that we have written replacing the weighted mean by summing the product of the community phylogenetic distance matrix and abundance products.

The other additional pairwise metrics that are frequently employed to calculate the phylogenetic diversity within a sample or community employ phylogenetic VCV matrices under a Brownian Motion trait evolution assumption instead of a phylogenetic distance matrix. This series of metrics were generated by Helmus et al. [70]. Recall that above we saw that the expected covariances (i.e., the off diagonal elements) in a phylogenetic VCV matrix are equivalent to the root-to-tip distance minus the one-half of the phylogenetic distance between two species. The first metric that Helmus et al. [70] produced is called the Phylogenetic Species Variability (PSV), which is expected to calculate the expected variance among species in a community phylogeny for a trait evolving under Brownian motion. This metric does not weight the expected variance by the abundances of the species and has been shown to be exactly half the mpd value when the phylogeny is ultrametric [71]. The second metric that Helmus et al. [70] produced is called Phylogenetic Species Evenness (PSE), which is identical to mpd.a when the phylogeny is ultrametric with the exception that the phylogeny is scaled from zero to one for PSE and not for mpd.a [71]. The final commonly used metric proposed by Helmus et al. [70] is called the Phylogenetic Species Richness, which is equivalent to multiplying mpd by the number of species in the community [71]. Thus, why conceptually different, the Helmus et al. metrics are essentially equivalent to metrics produced earlier or can be calculated by slightly modifying those metrics. It is important, though, to remember that this is only in the case when the phylogeny is ultrametric and that the framework presented by Helmus et al. is potentially more flexible because, in theory, one could implement phylogenetic VCV matrices that use alternative models of trait evolution to generate the expected variances and covariances. That said, the implementation of the Helmus et al. metrics in R currently does not permit alternative models of trait evolution. They can be calculated as follows.

```
> psv(my.sample, my.phylo)
> pse(my.sample, my.phylo)
> psr(my.sample, my.phylo)
```

We have now covered the list of the major pairwise metrics of PD that you will likely encounter in the literature. Undoubtedly, additional metrics exist and some are likely being generated this very instant, but in most cases these metrics will be conceptually and mathematically very similar to what is covered here. The challenge therefore is to determine when, where, and why metrics that will be produced in the future tell us additional meaningful biological information. In the next subsection we will consider nearest neighbor metrics of PD that are generally not related to pairwise metrics conceptually or mathematically.

3.5.2   Nearest Neighbor Measures

The use of pairwise distance measures is an effective means of calculating PD when one is interested in the overall diversity in a sample or community, but because these measures average across all species pairs a lot of detail is often lost. One important detail that is lost is "terminal" information. As I stated above, pairwise metrics are often considered "basal" because they average across all species, but these metrics cannot tell us the average phylogenetic distance for each species to its closest relative in the sample or community. It is often expected that the strongest interactions are likely to occur between closely related species [32, 72]. In such instances, we may be more interested in the "terminal" diversity. Similarly, ecologists investigating the trait diversity in communities frequently have calculated the average nearest neighbor in trait space across species in a community as a means to detect whether species are more broadly or finely spaced in one community compared to another. This concept lead to the development of mean nearest neighbor metrics for trait-based analyses that were translated to phylogenetics by Webb [28] again by simply replacing a trait distance matrix with a phylogenetic distance matrix. The unweighted nearest neighbor metric that Webb [28] produces is now generally referred to as the mean nearest taxon distance (*mntd*).

$$mntd = \frac{\sum_{i}^{n} \min \delta_{i,j}}{n}, \quad where\ i \neq j$$

where there are n species in the community, $\delta_{i,j}$ is the phylogenetic distance between species i and species j, and $\min\delta_{i,j}$ is the minimum phylogenetic distance between species i and all other species in the community (i.e., the nearest neighbor distance). As in the *mpd* calculation, the distance from one species to itself is not considered. If we consider the structure of a community phylogenetic distance matrix, it quickly becomes clear how to calculate *mntd*. In particular, for each row (i.e., species) in the matrix we need to find the smallest value that is not the diagonal value and then take a mean across rows. We can accomplish this with the following short function.

```
> new.mntd.function <- function(x){

    ## Get the names of the species present in a
    ## community.
    com.names <- names(x[x > 0])

    ## Make the community phylogenetic distance
    ## matrix by extracting those rows and columns
    ## that have species present in our community.
    my.com.dist <- dist.mat[com.names, com.names]

    ## Set all diagonal values to NA so that the
    ## zeros for conspecific comparisons do not
    ## interfere with our calculation of nearest
    ## neighbors.
    diag(my.com.dist) <- NA

    ## Use apply() to calculate the minimum value in
    ## each row of the community phylogenetic
    ## distance matrix and take a mean of those
    ## values.
    mean(apply(my.com.dist, MARGIN = 1, min), na.rm=T)

}
```

This small function for calculating the *mntd* for a single community can now be applied to all rows in the community data matrix.

```
> apply(my.sample, MARGIN = 1, new.mntd.function)
```

The *mntd* for all communities can also be calculated using the `mntd()` function in the *picante* package. Again the general workings of the code are similar aside from a `for()` loop through rows in the community data matrix rather than using an `apply()`.

```
> mntd(my.sample, cophenetic(my.phylo),
  abundance.weighted = F)
```

As with the tree-based and pairwise metrics described above, incorporating species abundance into the calculation of nearest neighbor metrics can likely yield some important insights. For example, two communities may both have very small *mntd* values indicating the co-occurrence of very closely related species, but one community could have nearest neighbors that have vastly different abundances whereas in the other community the abundances of the nearest neighbors may be similar. This information is critical for our understanding of when, where, and why very closely related species can co-occur. We can therefore derive an abundance-weighted version of the above nearest neighbor metric that we will call *mntd.a*.

$$mntd.a = \frac{\sum_{i}^{n} \min \delta_{i,j} f_i}{n}, \quad where\ i \neq j$$

where we have added the variable f_i to indicate the abundance of species i in the community. In order to calculate this new metric, we can see that we simply need to weight the mean by abundance of each species represented as rows in the community phylogenetic distance matrix. In other words, we must quantify the product of the abundance of a species with the minimum value found in the row in the community phylogenetic distance matrix for that species and take an average across all species. This can be accomplished with the following function.

```
> new.mntd.a.function <- function(x){

      ## Get the names of species present in the
      ## community.
      com.names <- names(x[x > 0])

      ## Make the community phylogenetic distance
      ## matrix using the names of the species present
      ## in the community.
      my.com.dist <- dist.mat[com.names, com.names]

      ## Place NA values in the diagonals
      diag(my.com.dist) <- NA

      ## Calculate a mean of the minimum values in each
      ## row of the community phylogenetic distance
      ## matrix weighed by the abundances of the
      ## species present in the community.
      weighted.mean(apply(my.com.dist, 1, min, na.rm =
      T), x[x > 0])

}
```

This function can now be applied to all rows in our community data matrix to calculate the *mntd.a* for all communities simultaneously.

```
> apply(my.sample, MARGIN = 1, new.mntd.a.function)
```

The *mntd.a* can be calculated using the preexisting mntd() function in the *picante* package with the abundance.weighted option flagged as true.

```
> mntd(my.sample, cophenetic(my.phylo),
abundance.weighted = T)
```

In general, the above two metrics are the primary nearest neighbor metrics of PD you will find in the literature. Interestingly, the phylogenetics literature has not adopted the version of these metrics prevalent in the trait literature that takes the standard deviation of the nearest neighbor values instead of the mean. Quantifying the standard deviation of the nearest neighbor values can be useful in determining whether the relative spacing in trait or phylogenetic distance is relatively homogeneous or heterogeneous among species. I know of no existing functions in R to calculate these metrics for phylogenetics, so we will write our own here. You will find they are essentially the same with a minor modification. We will call the standard deviation version of the *mntd* and *mntd.a* metrics sntd and sntd.a, respectively. The sntd value can be calculated for all communities as follows:

```
> sntd.function <- function(x){

    ## Get the names of the species present in a
    ## community.
    com.names <- names(x[x > 0])

    ## Make the community phylogenetic distance
    ## matrix by extracting those rows and columns
    ## that have species present in our community.
    my.com.dist <- dist.mat[com.names, com.names]

    ## Set all diagonal values to NA so that the
    ## zeros for conspecific comparisons do not
    ## interfere with our calculation of nearest
    ## neighbors.
    diag(my.com.dist) <- NA

    ## Use apply() to calculate the minimum value in
    ## each row of the community phylogenetic
    ## distance matrix and take a mean of those
    ## values.
    sd(apply(my.com.dist, MARGIN = 1, min), na.rm=T)

}

> apply(my.sample, MARGIN = 1, sntd.function)
```

The *sntd.a* value for all communities can be calculated but requires the computation of a weighted standard deviation. A weighted standard deviation can be calculated using the *SDMTools* package in R.

```
> install.packages("SDMTools")

> library(SDMTools)
```

Now that we have installed and loaded the *SDMTools* package, we can proceed to calculating the *sntd.a* values for all communities.

```
> sntd.a.function <- function(x){

    ## Get the names of species present in the
    ## community.
    com.names <- names(x[x > 0])

    ## Make the community phylogenetic distance
    ## matrix using the names of the species present
    ## in the community.
    my.com.dist <- dist.mat[com.names, com.names]

    ## Place NA values in the diagonals
    diag(my.com.dist) = NA

    ## Calculate a mean of the minimum values in each
    ## row of the community phylogenetic distance
    ## matrix weighed by the abundances of the
    ## species present in the community.
    wt.sd(apply(my.com.dist, 1, min, na.rm = T),
    x[x > 0])
}

> apply(my.sample, MARGIN = 1, sntd.a.function)
```

We have now covered the major tree-based and distance-based metrics of PD. As stated in numerous places above, many other metrics exist or may be published in the near future. It is my guess that most of these metrics will likely fall into one of the three general categories above and will likely be highly correlated with existing metrics in those categories. By covering the major metric in each category and breaking apart how they are calculated in R, it is my hope that you will be able to understand exactly what is being calculated in these metrics and in the code for other metrics you may come across.

3.6 Comparing Metrics

The labyrinth of phylogenetic and functional diversity metrics is vast and can be confusing. It is not unusual for a researcher to be asked to utilize a different metric of PD for their study at the request of a collaborator or reviewer. It is also common to find one to several "new" PD metrics published in the literature each year. The problem with this is that many of the metrics for PD that are now used are highly correlated and in some cases mathematically identical. This is bound to happen when a literature explodes, such as the phylogenetic community ecology literature, but we should also strive to recognize redundancy and understand which metrics are monotonic and which are not. We should also force ourselves to compare all of our "new" metrics against the existing list to assure that we are not publishing a "new" metric that has been around for years or a metric that has an incredibly subtle mathematical difference that still results it in being monotonic with an existing metric. Once the "cat is out of the bag" it is hard to eradicate a newly branded and wholly redundant metric from the literature, so we can at minimum here document which metrics are highly correlated. This should help you understand why your use of the mpd metric, the PSV metric, or Rao's pairwise metric simply does not matter. In the following lines of code we recalculate all metrics on the same dataset, plot them against one another, and calculate a Pearson's correlation for each combination (Fig. 3.5).

```
> richness <- rowSums(decostand(my.sample, method =
"pa", MARGIN = 1))

> pd.no.root <- pd(my.sample, my.phylo, include.root =
F)[,1]

> pd.root <- pd(my.sample, my.phylo, include.root =
F)[,1]

> pw <- mpd(my.sample, cophenetic(my.phylo),
abundance.weighted = F)

> pw.prime <- mpd(my.sample, cophenetic(my.phylo),
abundance.weighted = T)

> Rao.pw <- raoD(my.sample, my.phylo)$Dkk

> helmus.psv <- psv(my.sample, my.phylo)[,1]
```

```
> helmus.pse <- pse(my.sample, my.phylo)[,1]

> helmus.psr <- psr(my.sample, my.phylo)[,1]

> nn <- mntd(my.sample, cophenetic(my.phylo),
abundance.weighted = F)

> nn.prime <- mntd(my.sample, cophenetic(my.phylo),
abundance.weighted = T)

> outputs <- as.data.frame(cbind(richness, pd.no.root,
pd.root, pw, pw.prime, Rao.pw, helmus.psv, helmus.pse,
helmus.psr, nn, nn.prime))

> names(outputs) <- c("Richness", "PD.No.Root",
"PD.Root", "MPD", "MPD.abund", "Rao", "Helmus.PSV",
"Helmus.PSE", "Helmus.PSR", "MNTD", "MNTD.abund" )

> plot(outputs, pch = 16)
```

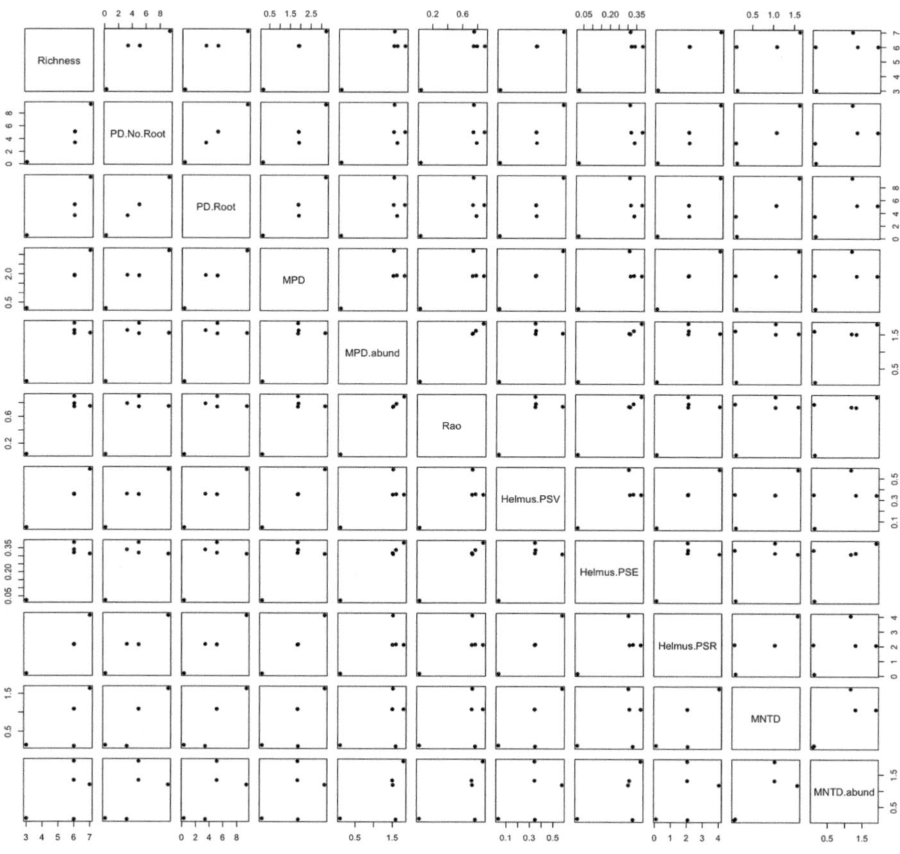

Fig. 3.5 A plot of each phylogenetic diversity metric against all other metrics. You will note that while some metrics are not related others are highly correlated in this dataset. For example, the abundance weighted mpd calculation, Rao and Helmus' PSE is all highly correlated and represent redundant metrics

```
> cor(outputs)
```

Although there is a lot of output to consider here, we can clearly see that some metrics are incredibly correlated if not exactly the same. This should serve as fair warning that the cottage industry of creating "new" metrics of PD should be scrutinized more carefully and we should not take any newly published metric as new or better. Indeed, in many cases that metric may be so similar to an existing metric that it is pointless to publish it. To reduce clutter I encourage those reviewing and generating new metrics to, at minimum, test whether the new metric is weakly correlated with an existing metric and use a simulation test (not a simple cooked example) to demonstrate that the new metric can pick up a biological process that other exiting metrics cannot detect.

3.7   Conclusions

This chapter has focused on quantifying the phylogenetic diversity of assemblages or communities. We have classified the metrics into tree-based metrics that are focused on summing the total phylogenetic branch length contained in a community and those that utilize a phylogenetic distance matrix to quantify the phylogenetic similarity of co-occurring species. Both approaches can be expanded to incorporate abundance information and by contrasting the equally weighted and abundance-weighted values, we may be able to have a better understanding of the phylogenetic distribution of abundance in a community. It is critical to remember that phylogenetic diversity metrics themselves tell us about just that and do not necessarily relate to the functional diversity of assemblage or community. That is, although ecologists have frequently used phylogenetic diversity in the past as a proxy of the ecological or functional similarity of species in the assemblage being studied, this usage is generally discouraged at this point in time because relatedness is not always a strong proxy of ecological similarity. Thus, phylogenetic diversity is still useful as a measure of biodiversity, but only in certain contexts. It is useful as an alternative indicator of biodiversity to inform managers and conservationists. It is also a useful alternative biodiversity metric that can be contrasted to functional and species diversity as independent variables in a statistical model and in those cases phylogenetic diversity is a stronger predictor of the response variable; this is evidence that unmeasured traits that may have phylogenetic signal are critical. Lastly, phylogenetic diversity may be useful in those cases where measuring function in the near future is not feasible and a quick measure of diversity beyond species richness is necessary. In any of these cases it will be critical for the researcher to understand whether or not the phylogenetic diversity metric is necessarily correlated with species richness, whether using abundance-weighting is necessary or informative, whether contrasting metrics is informative, and whether a null model is necessary to properly understand the phylogenetic pattern. Most of these issues can be addressed rather easily in R using the code provided in this and other chapters in the book.

3.8 Exercises

1. Simulate a coalescent phylogenetic tree using the `rcoal()` function in the *ape* package with the same number of species as in your example community data matrix from this chapter.
2. Assign names to the phylogeny that match those in the community data matrix and are in the same order.
3. Calculate the mean pairwise phylogenetic distance weighted by abundance and your simulated phylogeny.
4. Write a function that repeats steps 1–3 above 55 times. Take the resulting 55 vectors of mean pairwise phylogenetic distances and compile them into a matrix where the number of columns is 55 and the number of rows is equal to the number of communities in your example community data matrix.
5. Repeat steps 1–4 above, but use the `pd()` function to calculate species richness and Faith's Index. This time produce two matrices. In the first report the species richness for each community for each simulated phylogeny. In the second report the Faith's Index for each community for each simulated phylogeny.

Chapter 4
Functional Diversity

4.1 Objectives

The objectives of this chapter are to explore the variety of metrics and approaches for analyzing the functional composition and diversity of species assemblages. Important topics will include the consideration of how uni- and multivariate trait data are utilized in functional diversity analyses, the use of raw trait distance matrices versus trait dendrograms, and the degree of similarity between functional diversity metrics.

4.2 Background

The number of articles in ecology that are taking a "trait-based" approach is currently exploding with many of these articles seeking to quantify the functional diversity of the species in a community or assemblage (e.g., [73–92]). As we will see shortly, functional diversity can be quantified in a number of ways, but we can coarsely define it here as the diversity or dissimilarity of the ecological strategies or performance of species upon the basis of their morphological physiological traits. Traits directly or indirectly correlated with species performance (i.e., growth, mortality, reproduction) are increasingly termed "functional traits" and I will tend to use that nomenclature in the following text.

Despite the recent surge in interest, the measurement of functional diversity in communities or assemblages dates back at least 50 years with some of the most interesting early examples investigating the volume and packing of trait space of species in assemblages spanning an environmental or richness gradient [94–97]. For example, early work by Ricklefs and colleagues [93, 94] investigated whether the

The online version of this chapter (doi: 10.1007/978-1-4614-9542-0_4) contains supplementary material, which is available to authorized users

N.G. Swenson, *Functional and Phylogenetic Ecology in R*, Use R!,
DOI 10.1007/978-1-4614-9542-0_4, © Springer Science+Business Media New York 2014

spacing of species in trait space was maintained across a species richness as expected by limiting similarity theory [39].

The recent explosion of functional diversity measurement in ecology can largely be traced to influential work linking functional diversity to ecosystem function (e.g., [11, 91, 98]) and detailed reviews formalizing the conceptual foundation for functional diversity research (e.g., [76]). This work also coincided with the focus of plant ecologists on identifying a series of plant traits that are believed to be related to plant performance across environmental gradients [99–102]. Thus, in the span of two decades plant ecologists, in particular, found themselves collecting large and detailed datasets of plant traits in assemblages and developing and implementing measures of functional diversity. In many instances this has unfortunately resulted in the "reinvention" of metrics that were originally developed decades earlier and presumably forgotten or ignored. The goal of the present chapter is to discuss the main approaches for measuring functional diversity in communities and assemblages and how to calculate these metrics using R. We will not cover every possible metric of functional diversity ever published. Such an approach would be difficult and would result in the covering of many redundant measures. I have therefore chosen to keep it simple by covering the main classes of functional diversity metrics that are flexible. From each of these classes I will provide one or a few example metrics that are likely monotonic with many of the metrics you will encounter in the literature. As in the other chapters we will not simply learn how to "plug and chug" using existing functions. Rather we will dissect the measures so we understand how they work and so you may learn how you may construct your own measures of functional diversity or adjust those measures covered presently.

4.3 Quantifying the Functional Composition of Communities Using the Moments of Trait Distributions

As we will see in the following subsections, a great number of metrics have been generated for quantifying the functional diversity of communities with, often redundant, metrics invented nearly monthly at this point in time. It can be difficult to decipher what the results of various metric actually mean and how they do or do not relate to results published using alternative metrics. Outside of this increasingly complex maze of functional diversity metrics are simple calculations of the four moments of the trait distribution within communities or assemblages. The four moments—mean, standard deviation, skew, and kurtosis—are easier to interpret for the average scientist and can therefore be the easiest way to begin understanding the functional composition of your study system. They may even be an effective method for detecting the imprint of deterministic community assembly processes (e.g., [77, 84, 103]). For example, the co-occurrence of functionally similar species will be reflected by a lower standard deviation and a higher kurtosis of trait values in an assemblage [77].

In this section we will compute the four moments of the trait distribution in assemblages weighting all species present equally. We will then compute the community-weighted mean trait value for communities. The community-weighted mean is simply the mean trait value weighted by the relative abundance of species and it is becoming frequently used in trait-based ecological analyses. We begin by reading in the example community dataset for this chapter, which is in the format of a community data matrix. Recall that the community data matrix is the general format for most of the ecology-specific functions you will encounter in R. The example community data matrix can be read in as a table.

```
> my.sample <- read.table("FD.example.sample.txt",
header = T, row.names = 1)
```

Ensure that your data has been read into R correctly with the community (or site) names as row names, the species (or taxa) names as column names, and number of individuals as the cell values. To do this, take a quick look at the matrix.

```
> my.sample
```

Next we can read in the example trait data for this chapter. The .txt file has species names in the first column followed by four columns of trait data. It is important that the species names become row names when reading in the example or your own trait data into R for the analyses that follow.

```
> traits <- read.table("FD.traits.txt", sep = "\t",
header = T, row.names = 1)
```

Look at the trait data to ensure that the data were read into R correctly and that species names are indeed represented as row names and there are four columns of trait data.

```
> traits
```

The community data and trait data matrices have now been loaded into R and we are now ready to calculate the moments of the trait distributions for communities. To begin we will calculate individual moments for individual communities in order to understand the code. The first step is to quickly review how to get the list of species present in a community or assemblage. For example, we would like to know all of the species in our first community or assemblage with an abundance greater than zero—the first row in our community data matrix.

```
> spp <- names(my.sample[1, my.sample[1,] > 0])
```

The result is an object called "spp" that contains the names of species present in our first community. The trait data for these species can now be extracted from our traits matrix by asking for only those rows with the row names matching the names in our species list.

```
> traits[spp, ]
```

The above allows us to extract the trait data for all species in a community or assemblage for subsequent analyses. The two lines of code can also be combined into one line.

```
> traits[names(my.sample[1, my.sample[1,] > 0]),]
```

We now know how to get the values for all traits in our matrix for all species present in our first community or assemblage. From this output we can calculate the moments of the trait distribution. We can first calculate the community mean for each trait by wrapping the above line of code with the function mean().

```
> mean(traits[names(my.sample[1, my.sample[1,] >0), ],
na.rm = T)
```

The na.rm = T argument was utilized here in case your trait data matrix had missing trait values for some species. We can calculate the mean trait values for other communities or assemblages by altering what row is being selected from the community data matrix. For example, the next line of code calculates the mean trait values for community 3 instead of community 1.

```
> mean(traits[names(my.sample[3, my.sample[3,] >0), ],
na.rm = T)
```

At this point it is rather easy to calculate the remaining moments of the trait distribution. Here I calculate the standard deviation of the trait values in community 2.

```
> sd(traits[names(my.sample[2, my.sample[2,] >0), ],
na.rm = T)
```

High standard deviation values are indicative of more functional diversity in a community or assemblage, but they may be biased due to differences in the mean from community to community. To reduce this bias, a coefficient of variation in trait values can be calculated by dividing the standard deviations of the trait values in a community by the mean trait values in a community. This is easily done by dividing the last line of code by the line prior to it.

The functions mean() and sd() are in the *base* R package and are therefore available upon opening R. Functions to calculate skew and kurtosis, on the other hand, are not in the *base* package, but they are available in the *fBasics* package which can be installed and loaded as follows.

```
> install.packages("fBasics", dependencies = T)

> library(fBasics)
```

The skew of the trait distribution for community 1 can now be calculated in a similar way except using the function skewness().

```
> skewness(traits[names(my.sample[1, my.sample[1,]
>0), ], method = "moment", na.rm = T)
```

High values of skewness do not necessarily imply lower community functional diversity, but do indicate that most co-occurring species tend to have very similar trait values. The kurtosis of the traits in community 1 can be calculated using the `kurtosis()` function.

```
> kurtosis(traits[names(my.sample[1, my.sample[1,]
      > 0), ], method = "moment", na.rm = T)
```

Small kurtosis values indicate community trait distributions with "fatter" tails and therefore may indicate an increase in the average trait disparity between co-occurring species [77].

The above code calculates the moments of the trait distribution for one community, or row in the community data matrix, at a time. To automate this calculation across several communities (i.e., all rows in the community data matrix) we write a function where we calculate the moment for a single community and use an `apply()` function to apply the calculation to all rows (i.e., communities) in our system. In this example, we will write the moments for trait one in each community. We start by writing the kurtosis function for trait one.

```
> kurt.funk <- function(x){

    ## Calculate kurtosis for the first column in the
    ## trait matrix for species that are present in
    ## our community.
    kurtosis(traits[names(x[x > 0]), 1], method =
    "moment", na.rm = T)

}
```

The above code for calculating the kurtosis of the first trait in all communities can be easily changed to calculate the skewness and the mean and standard deviation as follows.

```
> skew.funk <- function(x){

    skewness(traits[names(x[x >0]), 1], method =
    "moment", na.rm = T)

}


> mean.funk <- function(x){

    mean(traits[names(x[x >0]), 1], na.rm = T)

}

> sd.funk <- function(x){

    sd(traits[names(x[x >0]), 1], na.rm = T)

}
```

These functions can now be applied to our community data matrix to calculate the mean, standard deviation, skew, and kurtosis of the trait values for the species present in each of our communities using the `apply()` function.

```
> apply(my.sample, MARGIN = 1, mean.funk)

> apply(my.sample, MARGIN = 1, sd.funk)

> apply(my.sample, MARGIN = 1, skew.funk)

> apply(my.sample, MARGIN = 1, kurt.funk)
```

In the above we have calculated the moments of the trait distributions in communities weighting all species or taxa present in the community equally. In many ecological analyses it is best not to treat all co-occurring species equally. Depending on the question of interest, it may be best to weight species by their abundances or some other measure of their dominance (e.g., percent canopy cover). One of the most common ways this is done when characterizing the functional composition of communities is to calculate what is has been termed the community-weighted mean (CWM) for a trait. The CWM is the mean trait value weighted by the relative abundance of each species. Here we will calculate the CWM for our communities using the same example datasets as before for comparison.

The first step in calculating the CWM is to transform our community data matrix from a tally of individuals of each species in a community into their relative abundances. These relative abundances can then be used to weight the mean trait value in a community. A rapid way to calculate the relative abundances of each species in each community is to divide the abundance of each species in a row (i.e., a community), by the total abundance in that row (i.e., a community). This can be accomplished using the function `rowSums()`.

```
> my.ra.sample <- my.sample / rowSums(my.sample)
```

All values in the cells of the matrix should now represent the fraction of the individuals in a community that for a particular species.

```
> my.ra.sample
```

It is possible that your data, for example, could contain the total canopy area or total biomass. In that instance the above code would provide you with percent cover or percent biomass. The R package *vegan* contains a function that can also be utilized to convert your community data matrix values into relative abundances, percent cover, or percent biomass. The function is called `decostand()` and can be used as follows.

```
> library(vegan)

> my.ra.sample2 <- decostand(my.sample, method =
"total", MARGIN = 1)
```

This function takes a community data matrix as input and can transform the matrix using several different methods applied to rows or columns. Here we have selected the "total" method and MARGIN=1, indicating that we want the method applied to each row. The method "total" divides each value in a row by the row sum or row total. Although this is what we did previously simply using the `rowSums()` function, the `decostand()` function is useful to know as many methods can be invoked easily. These methods include transforming the matrix to presence/absence (method="pa"), standardizing the values to a mean of zero and unit variance (method="standardize"), standardizing the values to range between zero and one (method="range"), and dividing the values by the maximum value in the row or column (method="max"). Thus, it is a powerful tool for altering the format of your community data matrix in a variety of ways. Now let us quickly check to see that our results using `rowSums()` are similar to that we received using `decostand()`.

```
> my.ra.sample2
```

You will find that the results of this approach and the previous approach are identical. To check that our relative abundances or percent cover or biomass for each community does indeed sum to one we can check the sum of each row.

```
> rowSums(my.ra.sample2)
```

Now that we have successfully changed our community data matrix to represent the relative abundances of species we can easily calculate the CWM for a single trait for a single community using the `weighted.mean()` function. This function takes an input matrix of values and an input matrix of weights. The input matrix of values in this case will be the trait values for the first trait of all species in our study system sorted using the order of the column names in the community data matrix.

```
> weighted.mean(traits[colnames(my.ra.sample), 1] ,
my.ra.sample[1, ])
```

The above selected the first column of the `traits` object to give the CWM of the first trait. The weighted mean of the second trait could be calculated by changing the selected column to 2.

```
> weighted.mean(traits[colnames(my.ra.sample), 2] ,
my.ra.sample[1, ])
```

To calculate the CWM value for the second trait in community two, simply change the row selected from the `my.ra.sample` object to 2.

```
> weighted.mean(traits[colnames(my.ra.sample), 2] ,
my.ra.sample[2, ])
```

Because the above code calculates the weighted mean for one trait in one community at a time, it is best that we write some code that will automatically calculate the CWM for a trait in all communities. Thus, we would like to apply our code to all communities simultaneously and will start by writing a function to calculate the weighted mean of trait one.

```
> weight.mean.funk <- function(x){
      weighted.mean(traits[names(x[x>0]), 1] , x[x > 0])
}
```

Now we can apply this function to our community data matrix to calculate the CWM for trait one in each community.

```
> apply(my.sample, MARGIN = 1, weight.mean.funk)
```

This approach can be extended to a weighted standard deviation as well using the `wt.sd()` function in the *SDMtools* package.

```
> library(SDMTools)
> weight.sd.funk <- function(x){
      wt.sd(traits[names(x[x>0]), 1] , x[x>0])
}
> apply(my.sample, MARGIN = 1, weight.sd.funk)
```

We have now see how to calculate the moments of the trait distribution for individual traits and individual communities. Some of these moments such as the mean cannot be deemed a measure of trait diversity. Indeed the mean is the exact opposite. Further, we may be interested in multivariate analyses of the function in a community. In the next sections we will discuss more detailed metrics designed to measure FD using one to many traits.

4.4   Dendrogram-Based Versus Euclidean Distance-Based Measures of Functional Diversity

The metrics for FD that we will discuss in the following all rely on a branch length or Euclidean distance to be measured between species. The branch length information comes from a dendrogram generated with a method of hierarchical clustering with a Euclidean trait distance matrix as input. This involves the clustering of species in trait space and may remove fine-scale trait differentiation between species. An alternative approach to this is to simply use the original trait distance matrix in the FD calculations. The use of the original distance matrix is appealing since the data are not transformed, but dendrograms are still frequently used to calculate FD and are sometimes preferred because their data structure is similar to that of a phylogeny which may be used in a concurrent analysis of phylogenetic diversity to which the FD analysis will be compared. Thus we will discuss both approaches presently.

4.4.1 Generating Trait Distance Matrices

The majority of the functional diversity metrics currently utilized in the literature are distance-based measures. One to many traits can be used in these metrics. The distances themselves are calculated either using the Euclidean distance between species in "trait space" or the branch lengths separating species on a dendrogram generated by clustering species based on their proximity in a trait distance matrix. In this subsection we will be focus on the calculation of the Euclidean distance between species in trait space.

First we will produce a distance matrix for the second trait in our trait matrix. To assure that we have species names on the output distance matrix we first generate a matrix containing the data for our second trait and with the row names from the original trait matrix.

```
> trait.2 <- as.matrix(traits[,2])

> rownames(trait.2) = rownames(traits)

> trait.2
```

Using this new object containing only the data for trait 2 we can calculate the Euclidean distance between all species upon the basis of this single trait. This is accomplished using the `dist()` function and the "euclidean" method.

```
> my.dist.mat.2 <- dist(trait.2, method = "euclidean")
```

As we will see below this distance matrix can be used in many of the existing metrics of functional diversity available in R. Although many studies will only analyze the functional diversity of an assemblage using a distance matrix constructed from multiple traits, it is generally useful to investigate the diversity of individual traits in an assemblage. Indeed the majority of the papers that have investigated the diversity of individual traits have found that not all traits behave similarly across assemblages (e.g., [64, 74, 75, 77, 79]). That is, one trait might increase in diversity along an environmental gradient while another may decrease. Thus, analyzing individual traits may help "unpack" the overall functional diversity calculated from multiple traits.

A distance matrix can be generated using all trait data simultaneously also using the `dist()` function. In this case, the `traits` input matrix already has species names as row names and therefore does not need any transformation.

```
> dist(traits, method = "euclidean")
```

While it is simple to compute the Euclidean distance between all pairs of species based on multiple traits, this approach is generally not advised. This is because it is likely that the measured traits may covary and may be measured on vastly different scales. For example, it is common in plant functional ecology to measure many covarying leaf traits and one or a few orthogonal stem or seed traits. If, for example, a distance matrix were computed on these raw data the leaf traits, which effectively only represent one axis of function, would dominate the structure of the distance matrix. Further, if a trait is measured, such as seed mass, that spans several orders of magnitude, that trait will dominate the structure of the distance matrix. Both of these scenarios are undesirable. Thus, it is often best to first transform your data to approximate a normal distribution for each trait. Then the data should be scaled and used in a principal components analysis to eliminate trait redundancy. We will make our data are approximately normal by using a `log()` in this example and we will scale the data using the `scale()` and `apply()` functions.

```
> traits.scaled <- apply(log(traits), MARGIN = 2,
scale)
```

The result is that all the trait values in each column are scaled to approximately a mean of zero and unit variance. These data can now be used in a principal components analysis to eliminate trait redundancy and to produce orthogonal axes of function that can be used in a distance matrix calculation. The scaled trait data can be input into the `princomp()` function as follows.

```
> pc = princomp(traits.scaled)
```

The principal components (PC) analysis will produce one PC axis for each input column (i.e., trait). If all PC axes were then used as input in a distance matrix calculation the result would be no different than if the PC analysis was not performed at all. Thus, we must select the few axes that explain the vast majority of the variance in the scaled trait data. We can examine the proportion of the total variance explained by each axis by examining a summary of the object output by the `princomp()` function.

```
> summary(pc)
```

A good rule of thumb is to include the PC axes that explain over 90 % and perhaps even 95 % of the variation in the scaled trait data. In this example, the first three PC axes explain 94.7 % of the variation. To determine what traits are most heavily weighted on these axes we can examine the trait loadings.

```
> print(pc$loadings, cutoff = 0.001)
```

From this we can see that trait 1 and trait 2 most heavily influence the first PC axes and traits 4 and 5 most heavily influence the second and third PC axes, respectively.

The next objective is to extract where each species lands on the first three PC axes and to use this information to generate a distance matrix. First we can simply look at these values by printing the PC scores to the screen.

```
> print(pc$scores, cutoff = 0.001)
```

We see one column for each PC axis and one row per species in the same order as the rows names (i.e., species names) in the input scaled distance matrix. As we are only using the first three PC axes we will extract the PC scores from those axes and put them in a new matrix assigning row names from the original `traits` matrix.

```
> pc.scores <- pc$scores[,1:3]

> rownames(pc.scores) <- rownames(traits)

> pc.scores
```

This data can now be used in the `dist()` function to calculate the multivariate Euclidean distance between all species in the study system.

```
> pc.dist.mat <- dist(pc.scores, method = "euclidean")

> pc.dist.mat
```

The result of this analysis is a distance matrix that is less likely to be biased by the co-variation of traits in the original dataset and differences in the scale of measurement between traits. It is therefore recommended that this approach be used in most cases for calculating a distance-based functional diversity metric. Of course, calculating the diversity of an actual trait and not PC scores is more intuitive to many. I do not completely discourage such an approach and I find it useful, but the individual trait diversities should not be assumed to be independent. Further, I do not recommend calculating a distance matrix using all raw trait data simultaneously given that many of the traits ecologists measure strongly covary. Similar to raw trait data it is often useful to investigate each individual axis of function. In the case of data that has been normalized, scaled, and run through a PC analysis, scores from individual PC axes can be used as individual "traits" for analysis.

The above assumes that the trait data you are utilizing is continuous data with no values missing for any traits or any species in your system. This is not always the case in ecological studies where rare species can often not be located in the field or an available database and where we are interested in using categorical traits such as growth form along with continuous traits. In instances where we have mixed trait variables (i.e., continuous and categorical) and/or a few missing trait values for a few species an alternative approach is necessary as a `dist()` function will not accommodate this scenario. A Gower Distance, though, can be calculated using the `gowdis()` function in the *FD* package. This function calculates the overall similarity of species or taxa based on Gower [104] and converts this similarity to a dissimilarity by subtracting it from one. If all traits are equally weighted, then the Gower similarity, for continuous variables, is the difference in values divided by the maximum observed difference for that variable in the dataset and this value is summed across all traits. For binary traits the contrast is either a zero or one for different or same. The Gower Distance for our dataset can be calculated as follows.

```
> library(FD)

> gowdis(traits)
```

Now that we have a trait distance matrix it can be used directly in FD analyses or as input into a hierarchical clustering analysis to generate a trait dendrogram. We will discuss trait dendrogram construction in the next subsection.

4.4.2 Generating Trait Dendrograms

The generation of trait dendrograms is quite simple. Perhaps it is so simple that the researcher often does not understand conceptually and mathematically what is being done in the background calculations. Thus, we will cover the few steps needed to calculate a trait dendrogram in R with some explanation of what is happening particularly with respect to the generation of the distance matrix and the clustering method.

The first step in generating a trait dendrogram is to produce a distance matrix representing the distance between all taxa or species in your system using one to many traits. In this example, we will first use multiple continuous traits stored in a matrix with species names as the row names. The most advisable and easily comprehended method for generating a trait distance matrix for construction of a dendrogram is to calculate the Euclidean distance between all species in trait space. Here we will assume that the traits are not correlated, but if your traits do covary please consult the code in the preceding section regarding reducing data redundancy prior to distance matrix generation.

```
> my.dist <- dist(traits, method = "euclidean")
```

As discussed above there may be instances where you are missing a trait value for a small number of species or you have mixed trait variables (i.e., continuous and categorical). In such instances the use of dist() is not possible and it is advisable to calculate a Gower distance using the gowdis() function in the *FD* package to produce a distance matrix for the next step. In this example, we will ignore such a situation.

We now use the trait distance matrix to generate a trait dendrogram using hierarchical clustering. The hclust() function in R performs hierarchical clustering using several different methods. In the majority of cases published in the literature an Unweighted Pair Group Method with Arithmetic Mean, commonly referred to as UPGMA, is utilized to generate a trait dendrogram. The UPGMA begins by identifying the two closest species in the trait distance matrix (randomly if there are multiple pairs sharing the smallest distance). A new distance matrix is then calculated between the distance between that cluster and all other species. The two species in this new matrix that are closest to one another form the next cluster and so on until all species and clusters are clustered. The branch lengths in the resulting dendrogram between clusters are calculated using pairwise distances.

$$cluster\,dist = \frac{\sum_i^A \sum_j^B d_{ij}}{A \times B}$$

where there are A species in cluster 1 and B species in cluster 2 and d_{ij} is the distance between all members of cluster 1 and members of cluster 2. Thus, the distances (i.e., branch lengths) between species in a trait dendrogram are no longer the distances between species in trait space. Rather, they are the distances between all the species in the clusters the species belong to in the dendrogram. A UPGMA-based dendrogram can be calculated using the `hclust()` function in R and the "average" method (Fig. 4.1).

```
> my.dendro <- hclust(my.dist, method = "average")

> plot(my.dendro)
```

We can see that the plotted dendrogram has branch lengths measured on a continuous scale corresponding to multivariate distance in trait space. In many instances it might be desirable to investigate a single trait with dendrogram-based metrics. Generating a dendrogram from a single trait (i.e., a single column in your trait matrix) is simple, but it is important to assign species names as row names. If they are not assigned the resulting dendrogram will not have species names on it. Here we will make a UPGMA dendrogram for the trait found in the second column of our trait matrix (Fig. 4.2).

```
> trait.2 <- as.matrix(traits[,2])

> rownames(trait.2) <- rownames(traits)

> my.dendro.2 <- hclust(dist(trait.2, method =
"euclidean"), method = "average")

> plot(my.dendro.2)
```

The plotted dendrogram shows that species do not cluster in the same way as they did previously when using all traits simultaneously. This indicates that species do not rank similarly on all trait axes and it highlights why it is often important and interesting to perform all analyses on all traits at once and each trait individually.

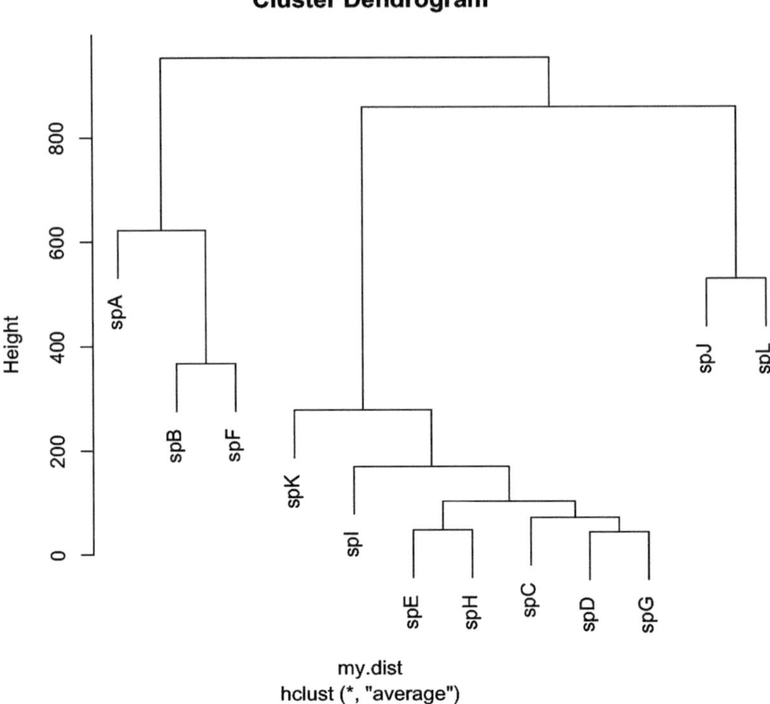

Fig. 4.1 A plot of our functional trait dendrogram constructed using UPGMA hierarchical clustering of a Euclidean distance matrix of all traits

4.4.3   Pairwise and Nearest Neighbor Measures

We are now ready to calculate the main two classes of FD metrics. Exactly as is the case for phylogenetic diversity (PD) in the previous chapter, the two main classes are pairwise distance metrics and nearest neighbor metrics. Indeed, the calculations and code are generally identical in R. The main decision we have to make prior to calculating these metrics for FD is whether to use a dendrogram or a raw trait distance matrix as the following code in this section requires a distance matrix that can be generated from the dendrogram or is the raw distance matrix. We will not use the dendrogram for the following analyses, but if we did we would first convert the dendrogram to a distance matrix.

```
> dendro.dist.mat <- cophenetic(my.dendro)
```

Thus we have effectively input a trait distance matrix into a hierarchical clustering algorithm to generate a dendrogram only to convert that dendrogram into a new distance matrix that likely has less refined information. You can now perhaps see why many do not like using a dendrogram-based approach unless necessary.

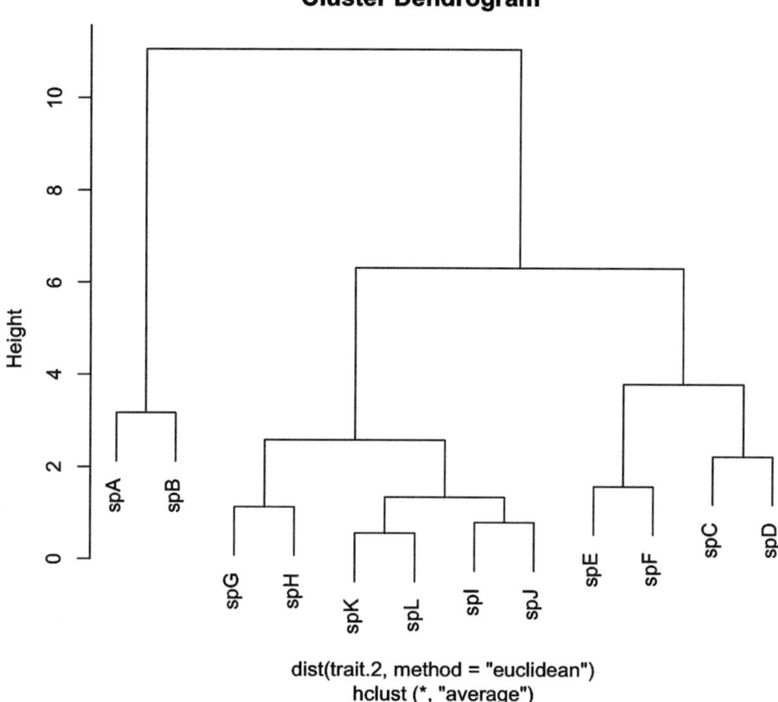

Fig. 4.2 A plot of our functional trait dendrogram for only our second trait constructed using UPGMA hierarchical clustering of a Euclidean distance matrix calculated for the second trait

For the remaining examples in this subsection we will use the raw trait distance matrix. If you recall above, we used `dist()` on our PC axes and we will now make this output a square matrix for the remaining functions.

```
> square.dist.mat <- as.matrix(pc.dist.mat)
```

The first FD measure we will compute is the unweighted pairwise functional distance between all present species in our second community. This can be calculated simply by first subsampling our distance matrix to only include present species and then calculating a mean value.

```
> mean(as.dist(square.dist.mat[names(my.sample[2,
my.sample[2,] > 0), names(my.sample[2, my.sample[2,] >
0)])])))
```

This can be calculated across all communities by first generating a small function to calculate the unweighted mean pairwise distance.

```
> trait.pw <- function(x){

    mean(as.dist(square.dist.mat[names(x[x > 0]),
    names(x[x > 0])])))

}
```

This function can now be applied to all rows (i.e., communities) in our community data matrix using the `apply()` function to compute the metric.

```
> apply(my.sample, MARGIN = 1, trait.pw)
```

This same calculation can be calculated using the `mpd()` function in the *picante* package which loops through communities rather than use an `apply()` making it computationally slower.

```
> mpd(my.sample, square.dist.mat, abundance.weighted =
F)
```

The abundance-weighted version of the pairwise trait distance metric can be calculated by again applying a function we will write. The function can be written as follows.

```
> trait.pw.prime <- function(x){

    ## Get the names of the species in the community.
    com.names <- names(x[x > 0])

    ## Make a matrix with one row containing
    ## abundances and names of all present species
    com <- t(as.matrix(x[x > 0]))

    ## Make functional distance matrix for taxa in
    ## community.
    com.dist <- square.dist.mat[com.names,com.names]

    ## Calculate the product of the abundances of all
    ## species in the community.
    abundance.products <- t(as.matrix(com[1,com[1, ] >
    0, drop = F]))%*% as.matrix(com[1,com[1, ] > 0,
    drop = F])

    ## Calculate a mean of the community functional
    ## distance matrix weighted by the products of
    ## all pairwise abundances.
    weighted.mean(com.dist, abundance.products)

}
```

The function is then applied to the rows of our community data matrix to calculate the abundance-weighted mean pairwise functional distance in each of our communities.

```
> apply(my.sample, MARGIN = 1, trait.pw.prime)
```

This metric can also be calculated using the `mpd()` function in the *picante* package.

```
> mpd(my.sample, square.dist.mat, abundance.weighted =
T)
```

The mean pairwise functional distances calculated above represent one major class of FD metric. Indeed, many of the "new" metrics that are published each year are monotonic with the mean pairwise functional distance.

The second major class of functional diversity metric uses nearest functional neighbor distances. There are two main ways that nearest neighbor distances have been utilized in the trait literature. The first, and most common, method is to take the mean nearest neighbor distance. In this calculation, the distance to the nearest functional neighbor in the community that is not your own species is tallied for each species and a mean is taken. This is easy to calculate but first we will place an NA value in the diagonal of our functional distance matrix to ensure we are not counting conspecifics in our nearest neighbor calculations.

```
> new.square.dist <- square.dist.mat

> diag(new.square.dist) <- NA
```

Now we can calculate the mean nearest functional neighbor distance for the species in our third community by first extracting a community-level distance matrix.

```
> com.dist.mat <- new.square.dist[names(my.sample[3,
my.sample[3,] > 0), names(my.sample[3, my.sample[3,] >
0)])])
```

From this distance matrix we can use the `apply()` function to calculate the minimum value in each row (i.e., the nearest neighbor for each species) and take a mean. This is the mean nearest neighbor distance for our community.

```
> mean(apply(com.dist.mat, MARGIN = 1, min, na.rm = T),
na.rm = T)
```

The second way nearest neighbor distances could be utilized in a study of FD is to take a standard deviation of the nearest neighbor values. This gives the researcher an idea of the regularity of the spacing and not just the average distance. A low variance indicates that species are relatively evenly placed in functional space.

```
> sd(apply(com.dist.mat, MARGIN = 1, min, na.rm = T),
na.rm = T)
```

We can apply both of these metrics to all rows (i.e., communities) in our study system by first writing a function for each to calculate the value for a single community. First the mean nearest functional neighbor distance can be coded as follows:

```
> trait.mnn <- function(x){

    ## Get the names of the species present in a
    ## community.
    com.names <- names(x[x > 0])

    ## Make the community functional distance
    ## matrix by extracting those rows and columns
    ## that have species present in our community.
    my.com.dist <- square.dist.mat [com.names,com.names]

    ## Set all diagonal values to NA so that the
    ## zeros for conspecific comparisons do not
    ## interfere with our calculation of nearest
    ## neighbors.
    diag(my.com.dist) <- NA

    ## Use apply() to calculate the minimum value in
    ## each row of the community functional
    ## distance matrix and take a mean of those
    ## values.
    mean(apply(my.com.dist, MARGIN = 1, min),na.rm = T)

}
```

Next, the function for the standard deviation of nearest neighbor distances.

```
> trait.sdnn <- function(x){

    ## Get the names of the species present in a
    ## community.
    com.names <- names(x[x > 0])

    ## Make the community functional distance
    ## matrix by extracting those rows and columns
    ## that have species present in our community.
    my.com.dist <- square.dist.mat [com.names,
    com.names]

    ## Set all diagonal values to NA so that the
    ## zeros for conspecific comparisons do not
    ## interfere with our calculation of nearest
    ## neighbors.
    diag(my.com.dist) <- NA

    ## Use apply() to calculate the minimum value in
    ## each row of the community functional
    ## distance matrix and take a mean of those
    ## values.
    sd(apply(my.com.dist, MARGIN = 1, min),na.rm=T)

}
```

They both can be applied to our dataset as:

```
> apply(my.sample, MARGIN = 1, trait.mnn)

> apply(my.sample, MARGIN = 1, trait.sdnn)
```

Both functions can also be extended to weight the mean or standard deviation by the abundance of the focal species. The functions would be coded and applied as:

```r
> trait.mnn.prime <- function(x){

    ## Get the names of the species present in a
    ## community.
    com.names <- names(x[x > 0])

    ## Make the community functional distance
    ## matrix by extracting those rows and columns
    ## that have species present in our community.
    my.com.dist <- square.dist.mat [com.names,com.names]

    ## Set all diagonal values to NA so that the
    ## zeros for conspecific comparisons do not
    ## interfere with our calculation of nearest
    ## neighbors.
    diag(my.com.dist) <- NA

    ## Use apply() to calculate the minimum value in
    ## each row of the community functional
    ## distance matrix and take a mean of those
    ## values.
    weighted.mean(apply(my.com.dist, MARGIN = 1, min),x[x >
0], na.rm = T)

}
```

Next, the function for the standard deviation of nearest neighbor distances.

```r
> trait.sdnn.prime <- function(x){

    ## Get the names of the species present in a
    ## community.
    com.names <- names(x[x > 0])

    ## Make the community functional distance
    ## matrix by extracting those rows and columns
    ## that have species present in our community.
    my.com.dist <- square.dist.mat [com.names,com.names]

    ## Set all diagonal values to NA so that the
    ## zeros for conspecific comparisons do not
    ## interfere with our calculation of nearest
    ## neighbors.
    diag(my.com.dist) <- NA

    ## Use apply() to calculate the minimum value in
    ## each row of the community functional
    ## distance matrix and take a mean of those
    ## values.
    wt.sd(apply(my.com.dist, MARGIN = 1, min), x[x >0],
na.rm = T)

}

> apply(my.sample, MARGIN = 1, trait.mnn.prime)

> apply(my.sample, MARGIN = 1, trait.sdnn.prime)
```

We have now successfully calculated the unweighted and abundance weighted of
the mean nearest functional neighbor distance and the standard deviation of the
nearest functional neighbor distances for each community in our system. There is

currently no function in an R package, that I know of, that calculates the standard deviation of the nearest neighbor distance, but the unweighted and abundance-weighted mean can be calculated using the `mntd()` function in the *picante* package.

```
> mntd(my.sample, square.dist.mat, abundance.weighted = F)

> mntd(my.sample, square.dist.mat, abundance.weighted = T)
```

Again, low values of the mean nearest functional neighbor distance indicate functionally similar species co-occurring or low FD, whereas high values indicate functionally dissimilar species are co-occurring and therefore high FD. In most cases, the mean nearest neighbor distance is sensitive to the species richness gradient across the communities being compared in your system. In particular, the mean nearest neighbor distance will decrease with increasing species richness in most cases. We will discuss the implications of this and null models in Chap. 6.

4.4.4   Ranges and Convex Hulls

One of the oldest and simplest measurements of FD has been to quantify the range of functions in a sample or community. In the case of a single trait this measure of FD is simply the range of the trait values in the community. When there are two or three measured traits the FD then becomes the area or volume of the two- or three-dimensional shape, respectively, representing the community trait space with vertices defined by the maximum and minimum value for each trait [90, 92, 94]. Calculation of these areas or volumes for the purpose of investigating the FD in communities started in the 1960s if not earlier [94], though the approach has been "invented" again recently (e.g., [90]) spurring an increasing number of papers using this conceptual approach. The methodological advance made during these reinventions is the possibility of calculating the trait volume for a community in more than three dimensions (i.e., using more than three traits). In particular, "convex hulls" can now be calculated for high-dimensional data and the volumes of these hulls can be quantified. The convex hull volume for a community is now commonly referred to as FRic or Functional Richness, though the term Functional Richness can often be confused or conflated with Functional Group Richness, which is a distinct measure [92]. The calculation of a convex hull or FRic is conceptually appealing because it can help a researcher understand how species pack and fill trait space. For example, the original use of such metrics in the 1960s and 1970s was to ask whether communities with more species have a greater morphological trait volume [93, 94]. This research was often done in the context of limiting similarity theory where the expectation was that the only way one could "add" species to a community was to add them to the periphery of trait space because adding them somewhere within the existing trait distribution would mean they would be too similar to invade. Further, researchers predicted that environments that are more abiotically benign and/or that have stronger biotic interactions will allow the invasion and success of peripheral phenotypes resulting in a large morphological volume whereas harsher abiotic environments will limit the invasion and success of peripheral phenotypes resulting in a small morphological volume.

The calculation of the convex hull volumes (i.e., FRic) in R is not difficult, but before we proceed to the measurement of FRic let us quickly determine how to calculate the univariate equivalent of calculating the range for each trait in a community. We begin by focusing on the first community in our community data matrix (i.e., row one) and extracting the names of all the species that are present in the community as determined by their positive values in the community data matrix.

```
> com.1.names <- names(my.sample[1, my.sample[1,] >
0])
```

We can now use the names of the species present in our first community to extract only those rows in the trait matrix containing the species in our fist community. This pruned trait matrix can then be used in an `apply()` function with a MARGIN of 2 to calculate the maximum trait value in each column (i.e., for each trait) and the minimum value. The difference between the outputs from these two `apply()` functions is the range for each trait in the first community.

```
> apply(traits[com.1.names, ], MARGIN = 2, max) -
apply(traits[com.1.names,], MARGIN = 2, min)
```

If we wanted to the above for our fourth community in a single line of code we would use the following.

```
> apply(traits[names(my.sample[4, my.sample[4,] > 0),
], MARGIN = 2, max) - apply(traits[names(my.sample[4,
my.sample[4,] > 0), ], MARGIN = 2, min)
```

Establishing how to calculate the ranges of all traits simultaneously for a single community now makes it clear how to scale the analysis up to analyze the range of all traits in all communities simultaneously. To accomplish this we first write a function that will calculate the range of all traits for a single community.

```
> range.function <- function(x){

    ## Get the names of the species present in our
    ## community.
    com.names <- names(x[x > 0])

    ## Calculate the range for each trait.
    apply(traits[com.names, ], MARGIN = 2, max) -
    apply(traits[com.names, ], MARGIN = 2, min)
}
```

We now apply this function to the rows in our community data matrix to calculate the range of all traits in all communities simultaneously.

```
> apply(my.sample, MARGIN = 1, range.function)
```

The calculation of trait ranges across communities can now be extended to calculate the overall convex hull volume (i.e., FRic). The calculations of convex hulls can now be done with multiple packages in R, but we will utilize the *geometry* package, which we can install and load using the following code.

```
> install.packages("geometry")

> library(geometry)
```

The function `convhulln()` in the *geometry* package can be used to perform all of the necessary calculations. We will begin by first quantifying the convex hull for only the first community in our community data matrix. The `convhulln()` function requires a matrix of continuous values that it will treat as the vertices where each column is a trait. Thus, we can use the following code to extract only those species present in our first community from the trait matrix and provide those values to the convex hull function.

```
> convhulln(traits[names(my.sample[1, my.sample[1, ]
    > 0), ])
```

We see that the function output a three-column matrix defining all of the vertices that constitute the convex hull, but no other information is provided. We must ask the function to output the volume of the convex hull choosing the "FA" option.

```
> hull.output <- convhulln(traits[names(my.sample[1,
my.sample[1,] > 0), ], options = "FA")

> names(hull.output)
```

A list has been output with three elements—hull, area, and vol. The hull element contains all of the vertices we saw above. The area element reports the area of the convex hull, but this is not of interest to us. The vol element contains the volume of the convex hull and therefore is our FRic value for this community.

```
> hull.output$vol
```

This approach for calculating the FRic for a single community can now be scaled up to calculate the FRic for all communities simultaneously using the following function.

```
> hull.function <- function(x){

    ## Get the names of the species present in our
    ## community.
    com.names <- names(x[x > 0])

    ## Calculate the hull using all traits.
    convhulln(traits[com.names, ], options =
    "FA")$vol

}
```

The hull function can now be applied to all rows in the community data matrix using an `apply()` function. This rapidly produces the convex hull volume or FRic for each community.

```
> apply(my.sample, MARGIN = 1, hull.function)
```

The FRic of the species in each community can also be performed by using the `dbFD()` function in the *FD* package. The *FD* package performs many additional useful analyses for trait-based ecology that we discuss in the next section. Here we will install and load the package.

```
> install.packages("FD")

> library(FD)
```

We will now use the `dbFD()` function to calculate FRic. This function calculates many metrics simultaneously, but for the moment we are only interested in outputting the FRic values for our communities.

```
> dbFD(traits, my.sample)$FRic
```

A valuable aspect of this function in the *FD* package is that it performs an initial test to determine whether the trait data require a reduction in dimensionality. This raises an important point that we have discussed above and that permeates through all functional diversity analyses. It is essential that redundant trait axes are reduced so as not to overly weight your metric of FD using several covarying traits. Fortunately, this function performs this for you, but you are advised that reducing dimensionality in other analyses is generally not done for you automatically.

4.4.5   Other Measures

The last three metrics that we will discuss are the Functional Evenness (FEve), Functional Divergence (FDiv), and Functional Dispersion (FDis). The FEve metric utilizes a minimum spanning tree (MST) to connect all species in functional trait space and measures the regularity of the species points along the branches of this tree and the regularity of their abundances. Thus, we may expect it to be very similar to a nearest neighbor metric. The FEve metric can be calculated as:

```
> dbFD(traits[colnames(my.sample), ], my.sample, w.abun = T,
stand.x = T )$FEve
```

The FDiv metric first measures the average distance of all species from the centroid of the trait space in a community and then sums the magnitude of the divergences from that mean. Thus higher values are supposed to indicate more dispersion towards the maximum and minimum of the range of traits. It can be calculated as

```
> dbFD(traits[colnames(my.sample), ], my.sample, w.abun = T,
stand.x = T )$FDiv
```

The FDis metric calculates the distance of each species from the centroid of the community traits. Thus, it is not quite the same calculation as the pairwise distance between species we can expect that it might be highly correlated. It is also similar to the FDiv metric, though conceptually perhaps easier to understand the biological meaning of FDis and it is likely a clearer indicator of trait dispersion in a community. The FDis can be calculated as:

```
> dbFD(traits[colnames(my.sample), ], my.sample, w.abun = T,
stand.x = T )$FDis
```

We have now calculated all of the major types and varieties of FD metrics that you will commonly see in the literature. In the next section we will quickly compare these metrics to evaluate their independence.

4.5   Comparing Metrics of Functional Diversity

As we consider diversity metrics in this book and elsewhere it is essential to consider their mathematical and statistical relationships in order to determine which are providing novel information and which are not and are simply monotonic with an existing metric. We have explored most of these metrics in the previous chapter on PD, but we will quickly plot them against each other and calculate their correlations (Fig. 4.3).

```
> richness <- rowSums(decostand(my.sample, method = "pa",
MARGIN = 1))

> pw <- mpd(my.sample, as.matrix(dist(traits)),
abundance.weighted = F)

> pw.prime <- mpd(my.sample, as.matrix(dist(traits)),
abundance.weighted = T)

> nn <- mntd(my.sample, as.matrix(dist(traits)),
abundance.weighted = F)

> nn.prime <- mntd(my.sample, as.matrix(dist(traits)),
abundance.weighted = T)

> fric <- dbFD(traits[colnames(my.sample), ],
my.sample, w.abun = T,   stand.x = T )$FRic

> feve <- dbFD(traits[colnames(my.sample), ],
my.sample, w.abun = T,   stand.x = T )$FEve

> fdiv <- dbFD(traits[colnames(my.sample), ],
my.sample, w.abun = T,   stand.x = T )$FDiv

> fdis <- dbFD(traits[colnames(my.sample), ],
my.sample, w.abun = T,   stand.x = T )$FDis

> outputs <- as.data.frame(cbind(richness, pw,
pw.prime, nn, nn.prime, fric, feve, fdiv, fdis))

> names(outputs) <- c("richness", "pw", "pw.prime",
"nn", "nn.prime", "FRic", "FEve", "FDiv", "Fdis" )

> plot(outputs, pch = 16)

> cor(outputs)
```

We see that the PW and FDis metrics are almost identical as has been previously pointed out by Laliberte and Legendre [92] and that FDis and FDiv, while being correlated, are not identical indices. Lastly, we see that the hull/FRic metric and the mean nearest neighbor metric are related to species richness, indicating that they should perhaps be considered in the context of a null model for comparative analyses (see Chap. 6).

4.6   Conclusions

This chapter has addressed a large number of metrics that can be used to characterize the functional structure and diversity of communities or assemblages. We began by simply characterizing the moments of the trait distribution in assemblages. While this is useful, these measures, particularly the mean, are not measures of functional diversity and should not be used as such. We then covered pairwise- and nearest

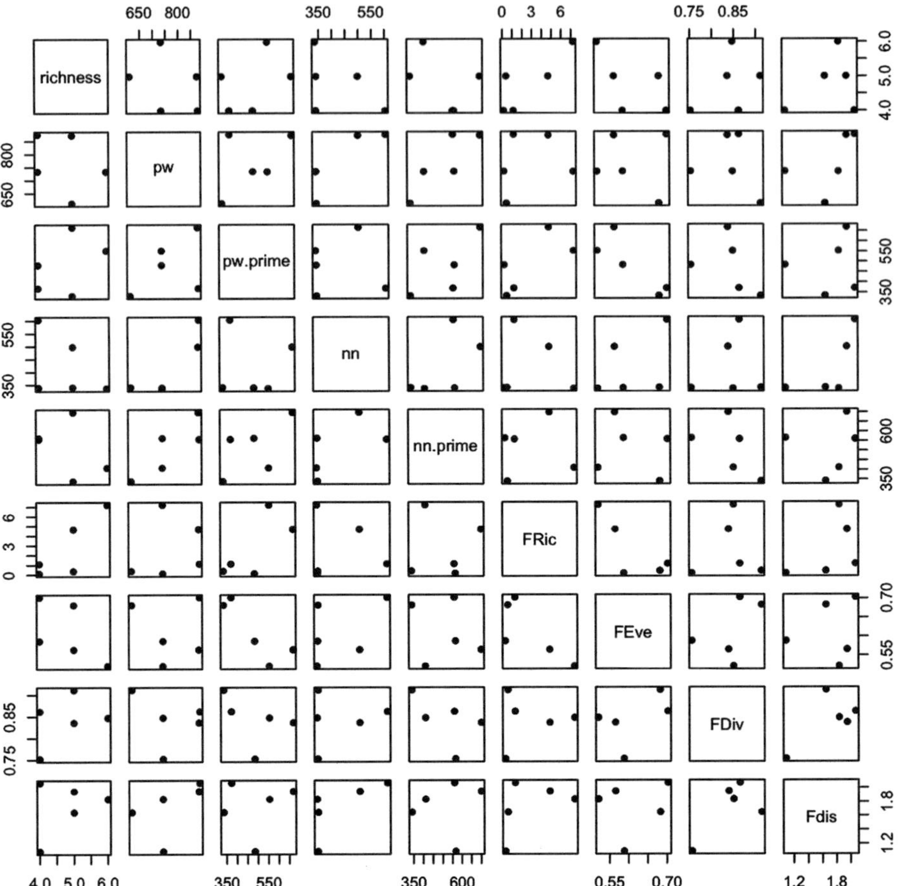

Fig. 4.3 A plot of each functional diversity metric against all other metrics

neighbor-based metrics of functional diversity which have their roots in the eco-morphology literature spanning back to at least the 1960s underscoring the long history of measuring trait similarity and functional diversity in species assemblages.

Almost all functional diversity metrics that have been published or that will be published will likely fall into either the pairwise or nearest neighbor classes. There are likely many metrics of functional diversity that I have not covered in this chapter, but it is likely that many of these are highly correlated with the measures we have covered. In deciding the "best" metric for your study I urge you to consider whether your chosen metric is actually that different from existing metrics. In many cases we may find that the metric is monotonic with an existing metric and may not provide much additional metric. It is also expected that the code above will provide you with enough details regarding how to handle trait and community data in R to formulate your own measures of functional diversity in R, but I again urge you to make sure your new metrics are indeed novel.

4.7 Exercises

1. Simulate two traits on your example phylogeny from Chap. 3 using Brownian Motion and the `fastBM()` function in the *phytools* package.
2. Quantify the functional diversity for each of your communities using the simulated trait data, the example community data matrix from Chap. 3, and the pairwise and nearest neighbor metrics weighted by abundance. Next do the same for phylogenetic diversity using the analogous pairwise and nearest neighbor metrics and the same phylogeny you used to simulate the trait data.
3. Perform a simple correlation between the phylogenetic and functional diversity measures. Do the functional and phylogenetic diversity measures correlate? Why or why not?
4. Repeat the above with other metrics of that are presence–absence weighted or weighted by abundance.

Chapter 5
Phylogenetic and Functional Beta Diversity

5.1 Objectives

The first objective of this chapter is to introduce the conceptual and empirical background for why phylogenetic and functional analyses of beta diversity are of interest to ecologists. This is followed by detailed instructions on how to calculate in R the major metrics of phylogenetic and functional beta diversity that are most commonly employed in the literature. The ultimate goal is to obtain a robust conceptual and practical knowledge of phylogenetic and functional beta diversity.

5.2 Background

The study of beta diversity has gone from being a topic in the backwaters of ecology to being an extremely hot and controversial topic in the literature. Even when and where we should use the simple term "beta diversity" has become amazingly controversial [105–107]. I therefore think it is worthwhile to begin with a brief statement on how I use the term "beta diversity." I am not of the camp that believes there is a "true" measure of beta diversity or that the term beta diversity should only be used in certain contexts any more than "biodiversity" should be defined by one and only one equation. I am of the opinion that we are wise enough to understand the general concept and to understand the metrics that are being utilized to represent this concept and how they differ from one another (see [108]). I therefore use the term beta diversity to represent the turnover in composition between a pair of assemblages or communities or the variation in the composition between a set of assemblages or communities. I purposefully use the term "composition" here because beta diversity should not be restricted to species names and their similarity between communities. Indeed, comparing the similarity of species names between

The online version of this chapter (doi: 10.1007/978-1-4614-9542-0_5) contains supplementary material, which is available to authorized users

N.G. Swenson, *Functional and Phylogenetic Ecology in R*, Use R!,
DOI 10.1007/978-1-4614-9542-0_5, © Springer Science+Business Media New York 2014

communities is very uninformative in many cases given that species are not statistically independent entities—they share differing degrees of evolutionary history and functional similarity [80, 89, 109–112]. Thus, in the following I will trust that you understand that beta diversity generally means, in my lexicon, the turnover or dissimilarity in the composition of assemblages or communities. If you do not appreciate this opinion you can simply replace the term "beta diversity" in the following with "dissimilarity" and proceed.

Now that we have dispatched with what exactly I mean by "beta diversity" we should consider why phylogenetic and functional information is interesting in the context of beta diversity. The argument for using phylogenetic and functional in beta diversity studies can be quite simple in that species are non-independent entities that share evolutionary history and vary in their functional similarity. If this general statistical argument is not enough, we can consider the evidence from recent research that has shown the turnover or variation in community composition from a "species" or "functional" perspective is often very different. For example, Fukami et al. [109] have shown that although the spatial turnover of the species composition of communities may diverge, the functional composition may converge. Further, Swenson et al. [110] have shown that although there is a strong temporal decay in the species compositional similarity in a tropical forest through time the functional composition does not turn over at all. Thus, in both cases the species beta diversity analyses are likely very misleading regarding the actual deterministic ecological process acting in the system. Such instances are likely particularly important in high diversity systems where many species are very similar or even redundant functionally and all that matters is which species with the "right" function can colonize the open site first.

Despite the conceptual arguments regarding the importance of measuring the phylogenetic and functional beta diversity between communities, the first clear measurements of phylogenetic beta diversity in the literature arose out of necessity in the microbial ecology literature [113, 114]. In particular, microbial ecologists are faced with utilizing phylogenetic and not Latin binomials to quantify the diversity in their study systems. Traditional species beta diversity metrics such as a Jaccard's distance or a Bray–Curtis distance were therefore of no use and metrics based on phylogenetic branch lengths were required. Microbial ecologists were therefore the first to generate tree-based and distance-based phylogenetic metrics of beta diversity, although this fairly deep literature is often ignored by ecologists working on plant and animal systems (e.g., [115]). Over the past 5 years plant and animal ecologists have become increasingly interested in phylogenetic and to some extent functional beta diversity and have generated their own metrics that are often highly related to the existing metrics published in the microbial literature.

In the following sections we will cover tree-based metrics of phylogenetic beta diversity that have been the focus of microbial ecologists. We will then transition to distance-based metrics of phylogenetic beta diversity metrics that can also be used for the measurement of functional beta diversity. These distance-based metrics are typically easier to understand conceptually, more flexible and faster to calculate, likely making them the metrics you will most frequently encounter in the literature. Although some of the metrics described, such as UniFrac, can be calculated using GUIs available online I would encourage you to use R for such analyses for transparency and flexibility.

Indeed some online programs do not even report the beta diversity statistic and only allow you to cluster communities or correlate the statistic with an environmental gradient making these programs useful for only a very narrow range of analyses.

5.3   Tree-Based Measures of Phylogenetic Beta Diversity

The measurement of phylogenetic and functional beta diversity in ecology is generally a new phenomenon and the literature on this topic does not have nearly the depth as that covering phylogenetic and functional alpha diversity. We will begin by discussing the two major tree-based metrics of phylogenetic and functional diversity. Generally, these metrics have only been used on phylogenetic trees, but could potentially be utilized on dendrograms constructed from trait data.

5.3.1   UniFrac

To the best of my knowledge, the first tree-based metric developed for computing phylogenetic or functional beta diversity was published by Luzopone and Knight [114]. Their metric seeks to quantify the unique fraction of the phylogeny contained in each of the two communities or assemblages being compared and hence the name UniFrac. The formal equation for calculating UniFrac is

$$UniFrac_{A,B} = \frac{PD_{A \cup B} - PD_{A \cap B}}{PD_{A \cup B}}$$

where $PD_{A \cup B}$ is the phylogenetic diversity of the species in communities A and B combined and $PD_{A \cap B}$ is the phylogenetic diversity of the species shared between communities A and B.

Put in plain English, in order to calculate the UniFrac for two communities you must subtract the shared branch lengths between two communities from the sum of the total branch lengths for all species in both communities (i.e., the Faith's index for a "community" containing all species found in your two communities). This difference is then divided by the sum of the total branch lengths in both communities. Thus, it is fairly easy to calculate in R. We start by first loading our example community data matrix for this chapter in to R.

```
> my.sample <- read.table("beta.example.sample.txt",
sep = "\t", row.names = 1, header = T)
```

Next we read in our example phylogenetic tree for this chapter.

```
> my.phylo <- read.tree("beta.example.phylo.txt")
```

To start we will simply calculate the UniFrac for the first two communities in our community data matrix. The first goal is to calculate the phylogenetic diversity (PD) using Faith's Index for the first and second communities independently. We can accomplish this by using only the first or second rows of the community data matrix in the pd() function. Given that the pd() function returns a PD value in the first column and a species richness (SR) value in the second column, we will ask for only the value in the first column.

```
> pd.1 <- pd(t(as.matrix(my.sample[1, ])), my.phylo)[,
1]

> pd.1

> pd.2 <- pd(t(as.matrix(my.sample[2, ])), my.phylo)[,
1]

> pd.2
```

The numerator and denominator of the UniFrac calculation require the PD calculated using all species found in both communities being compared. This can be accomplished by adding the two rows from the community data matrix containing the two communities being compared and using the resulting matrix in the pd() function.

```
> pd.total <-
pd(t(as.matrix(my.sample[1,
]))+t(as.matrix(my.sample[2, ])), my.phylo)[, 1]
```

The pd.total object that we just calculated describes the PD if the two communities were considered one. The branch length shared between the communities can be calculated by subtracting this value from the sum of the individual PD values for the two communities.

```
> pd.shared <- (pd.1 + pd2) - pd.total
```

The shared branch length can then be subtracted from the total branch length found when combining both communities to produce the total branch length that is unique to one community or the other (i.e., the numerator of the UniFrac equation). This value is standardized by the total branch length found when combining the two communities.

```
> u.frac <- (pd.total - pd.shared) / pd.total

> u.frac
```

The result is the fraction of the total branch length for the two communities that is unique to one of the two communities. The implementation described above is referred to as the "unweighted UniFrac" as it is not weighted by the abundance of individual species. The unweighted UniFrac can be calculated between all pairs of communities in the community data matrix using the unifrac() function in the *picante* package.

```
> library(picante)

> unifrac(my.sample, my.phylo)
```

It is often useful when calculating beta diversity to consider the relative abundances of the species in the communities being compared and to perhaps compare and contrast such measures to those that weight all species equally. The UniFrac metric is the only tree-based metric of which I am aware that has been extended to include the relative abundances of species into the calculation. Specifically, Luzopone et al. [116] extended their original unweighted UniFrac calculation to consider how many individuals from each community were subtended by the branches unique to each community. The calculation of the weighted UniFrac metric has two steps. The first step is to calculate a raw weighted UniFrac value (u) as follows:

$$u = \sum_{i}^{n} b_i \times \left| \frac{A_i}{A_\mathrm{T}} - \frac{B_i}{B_\mathrm{T}} \right|$$

where b_i is the length of individual branch i of the n total branches in your phylogeny containing all species in your community data matrix or meta-community. Each individual branch (b_i) is multiplied by the absolute value of the difference between the abundance of the species subtended by branch i in community A (A_i) divided by the total abundance in community A (A_T) and the abundance of the species subtended by branch i in community B (B_i) divided by the total abundance in community B (B_T). Simplified, the length of each branch is multiplied by the absolute value of the differences in the relative abundance of the species in communities A and B that are subtended by that branch summed over all branches. Given the calculation utilizes the relative abundances of species in each community we first transform our community data matrix to transform raw counts of species to relative abundances.

```
> my.ra.sample <- my.sample / rowSums(my.sample)
```

The next goal is to loop through each branch to calculate the product of its length and the absolute difference of the relative abundances of the species in the two communities subtended by that branch. We will start by generating an empty matrix that will hold the necessary information for the calculation where we will have one row per branch (i.e., edge) in our phylogeny and six columns.

```
> node.bls <- matrix(NA, nrow = length(my.phylo$edge.
length), ncol = 6)
```

Next we need a way to extract the basal and terminal node number for each branch and to link this information to the length of the branch. In the first two columns we place the numbers for the basal and terminal nodes of each branch.

```
> node.bls[,1:2] <- my.phylo$edge
```

In the first column are the basal node numbers for each branch in our phylogeny and in the second column are the terminal node numbers. To visualize these nodes first

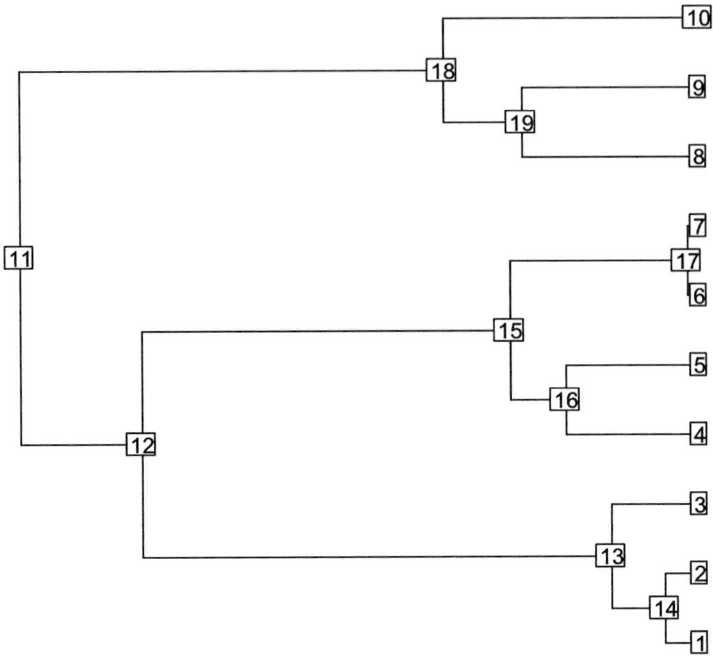

Fig. 5.1 A plot of our example phylogeny indicating the internal and terminal node numbers. Visualizing these node numbers helps to understand how to select and weight individual branches for the UniFrac calculations

plot the phylogeny using `plot.phylo()` and then label the internal node numbers using nodelabels() and the terminal node numbers using `tiplabels()` (Fig. 5.1).

```
> plot.phylo(my.phylo, show.tip.label = F)

> nodelabels(bg = "white")

> tiplabels(bg = "white")
```

The next objective is to match each of these node combinations with a branch length and placing the value in the third column.

```
> node.bls[,3] <- my.phylo$edge.length

> head(node.bls)
```

We now have a matrix containing the numbers indicating the placement of each branch in the phylogeny and the length of each of those branches. The length of the branches corresponds to the parameter b_i in the raw weighted UniFrac equation. We will now use a `for()` loop to loop through each row in the matrix identifying the leaves (i.e., taxa) subtended by each branch, calculating the absolute difference in their relative abundances in two communities, and multiply that by the branch length. The calculation will require the `tips()` function from the *geiger* package. This function reports the names of the taxa subtended by the node number provided. First we load the *geiger* package and then start the loop.

```
> library(geiger)

## Begin the loop from row 1 to the number of rows in
## the node.bls object
> for(i in 1:dim(node.bls)[1]){

    ## Using the terminal node number for each branch
    ## in column two of the node.bls object we pull
    ## out the taxa names subtended by the branch
    ## described in each row of the node.bls object.
    leaves.node <- tips(my.phylo, node.bls[i, 2])

    ## Sum the total relative abundance of the
    ## species found in community 1, which is the
    ## first row in our my.ra.sample object. This is
    ## equal to the Ai/AT in the raw weighted UniFrac
    ## equation.
    node.bls[i, 4] <- sum(my.ra.sample[1,
    leaves.node])

    ## Sum the total relative abundance of the
    ## species found in community 1, which is the
    ## first row in our my.ra.sample object. This is
    ## equal to the Bi/BT in the raw weighted UniFrac
    ## equation.
    node.bls[i, 5] <- sum(my.ra.sample[2,
    leaves.node])

    ## Calculate the product of the branch length in
    ## column 3, the bi parameter) and the absolute
    ## value of the difference of the relative
    ## abundances in the two communities for the
    ## species subtended by the branch.
    node.bls[i, 6] <- node.bls[i, 3] *
    (abs(node.bls[i, 4]-node.bls[i, 5]))

## close loop
}

## Calculate the raw weighted UniFrac by summing the
## values in column 6.
> raw.weighted.unifrac.output <- sum(node.bls[, 6])
> raw.weighted.unifrac.output
```

The result is the raw weighted UniFrac value (u) for the first two communities in our community data matrix. This value is typically "normalized" by a scaling factor called D. In words, the scaling factor sums the relative abundance for each species in the two communities and multiplies this value by the branch length from the root of the phylogeny to the branch tip for that species and sums this calculation over all species.

$$D = \sum_{j}^{n} d_j \times \left(\frac{A_j}{A_T} + \frac{B_j}{B_T} \right)$$

where d_j is the distance from the root to the branch tip for species j of n total species. The value for d_j will be constant for all species when using an ultrametric phylogeny. The abundances of species j in community A (A_j) and community B (B_j) are divided by the total abundance of all species in community A (A_T) and B (B_T).

In order to calculate the scaling factor, D, we can utilize a loop where we calculate the total relative abundance of each species in the two communities and multiply that value by the phylogenetic distance from the root to the tip holding each species. First we make an empty matrix to hold the three columns for the output of interest and one row per species in our community distance matrix. If the species is not present in either of our communities it is fine as the total relative abundance will be zero and will not influence the summation.

```
> scale.output <- matrix(NA, ncol = 3,
nrow=ncol(my.ra.sample))
```

Now start the loop from 1 to the number of species in our meta-community (i.e., the number of columns in our community data matrix object my.ra.sample).

```
## Start the loop
> for(j in 1:ncol(my.ra.sample)){

        ## add the relative abundance of species j in
        ## both communities. This is the parenthetical
        ## calculation in the scaling factor equation.
        scale.output[j, 1] <- my.ra.sample[1, j] +
        my.ra.sample[2, j]

        ## Use the diagonal values produced from the
        ## phylogenetic variance-covariance matrix
        ## function vcv.phylo() to provide the distance
        ## from the root to the tip holding species j.
        ## This is parameter dj.
        scale.output[j, 2] <-
        vcv.phylo(my.phylo)[colnames(my.ra.sample)[j],coln
        ames(my.ra.sample)[j]]

        ## Multiply the branch length, dj, by the total
        ## relative abundance for species j in the two
        ## communities.
        Scale.output[j, 3] <- output[j, 2] * output[j, 1]

    ## Close the loop.
    }

## Calculate the scaling factor, D, by summing the
## third column of the output.
> scaling.factor.output <- sum(scale.output[, 3])

> scaling.factor.output
```

We can normalize the raw weighted UniFrac, u, by dividing by the scaling factor, D.

```
> raw.weighted.unifrac.output / scaling.factor.output
```

We, therefore, now have the ability to calculate the ability to calculate a weighted and scaled version of the UniFrac metric. The weighting of UniFrac by abundance is a welcome improvement over the original UniFrac metric, but some conceptual and mathematical problems with this weighted metric have been pointed out [117]. One particularly troubling problem is that the weighted UniFrac does not equal the value from the original UniFrac calculation if abundances become binary. That is, if species are treated as either present or absent in the weighted equation the result would not equal that of the original equation. In other words, the weighted metric is not a natural extension of the original and the two are not coherent. An additional issue with the weighted UniFrac calculation above is that it may overly weight branches with large abundance proportions [117]. A solution to these two issues has been derived by Chen et al. [117] by altering the equation for weighted UniFrac to include a normalizing factor to the equation.

$$d^{(\alpha)} = \frac{\sum_{i}^{n} b_i \left(p_i^A + p_i^B \right)^\alpha \left| \dfrac{p_i^A - p_i^B}{p_i^A + p_i^B} \right|}{\sum_{i}^{n} b_i \left(p_i^A + p_i^B \right)^\alpha}$$

where α is the normalizing factor that can range from zero to one, b_i is the length of branch i, and p_i^A is the proportion of the abundances subtended by branch i for community A. This modified, and arguably much improved, version of the abundance-weighted UniFrac is available in the *GUniFrac* package.

```
> install.packages("GUniFrac")
```

```
> library(GUniFrac)
```

The GUniFrac() function in the *GUniFrac* package can run the analyses with multiple different levels of alpha at the same time to produce an array of dissimilarity matrices.

```
> GUniFrac(my.sample, my.phylo, alpha = c(0, 0.5, 1))
```

The first three layers of the output array are the weighted UniFrac values with the three different levels of alpha we provided. As the alpha parameter approaches one it places more weight on branches with more abundance. The fourth level is the unweighted version of the Chen et al.'s [117] UniFrac equation above and the last level is the variance-weighted version of UniFrac from Chang et al. [118]. An issue, of course, with the Chun et al.'s metric is that one must pick the results from an alpha level for downstream analyses. If this was a uniformed process it would be undesirable, though quantifying the sensitivity of the downstream inferences resulting from different alpha levels. Despite these drawbacks there are also clear drawbacks to the original weighted version of UniFrac, suggesting that more work is needed to understand the behavior of these metrics and it is perhaps best to utilize distance-based metrics until these issues are resolved.

5.3.2   Phylogenetic Sorenson's Index

The next metric we will discuss is very similar to the unweighted UniFrac method. It was proposed years later by Bryant et al. [119] and is calculated as

$$PhyloSor = 2 \times \frac{BL_{k_1 k_2}}{\left(BL_{k_1} + BL_{k_2} \right)}$$

where $BL_{k_1 k_2}$ is the Faith's Index of the species shared between two communities and BL_{k_1} and BL_{k_2} are the Faith's Indices for the two communities. The PhyloSor metric is a metric of similarity such that as the phylogenetic similarity of two communities increases the PhyloSor value increases. This is the opposite of the unweighted *UniFrac* metric which is a dissimilarity metric. Nonetheless, as we will see below, the two metrics are highly correlated.

We can calculate the PhyloSor for two communities using all the calculations we did in the previous subsection for the unweighted UniFrac. Specifically, all we need to do is to divide the Faith's Index for the species shared by the two communities by the sum of the Faith's Index for each of the two communities and multiply this value by 2.

```
> 2 * (pd.shared / (pd.1 + pd.2))
```

Thus we can see what PhyloSor is closely related to the unweighted UniFrac metric. Indeed, the two are monotonic and the PhyloSor metric represents essentially a reinvention of a preexisting phylogenetic beta diversity metric. Thus, there is really no quantitative reason to think that PhyloSor is an improvement over the preexisting unweighted UniFrac metric.

The PhyloSor of all communities can be calculated using the `phylosor()` function in the *picante* package. It should be noted that this calculation is incredibly slow such that it is potentially not useful when dealing with large phylogenetic trees and/or large numbers of communities.

```
> phylosor(my.sample, my.phylo)
```

We have now seen the two major types of tree-based phylogenetic beta diversity metrics. Conceptually both metrics could be used to calculate functional beta diversity, but this can be tricky in R as the existing code requires a phylo object. A functional dendrogram can be coerced into a phylo object using `as.phylo()`, but it may cause unwanted distortions to the dendrogram. The distance-based beta diversity approaches we will cover in the next section are amenable to both phylogenetic and functional data and are often much faster to compute than the above tree-based metrics.

5.4   Distance-Based Measures of Phylogenetic and Functional Beta Diversity

The above metrics of phylogenetic beta diversity, in essence, both compare the relative proportion of the total amount of Faith's Index that is shared or unique to both communities, thereby defining their similarity or dissimilarity. The calculations for these tree-based methods can be slow and the metric that has been modified to incorporate abundances has some outstanding issues regarding how it should or should not be used.

An alternative class of distance-based beta diversity metrics exist that are generally faster to compute, are able to handle phylogenetic and functional information, are more tractable when transitioning from presence–absence weighted to abundance weighted, and are conceptually and mathematically tightly linked with existing distance-based phylogenetic and functional alpha diversity metrics. These properties of distance-based beta diversity measures make them preferable for most analyses.

The following subsections will explore the two main groups of distance-based phylogenetic and functional beta diversity metrics. Similar to distance-based alpha diversity metrics, the two main groups are based on pairwise and nearest neighbor distances. Recent work by Swenson [112] has demonstrated that all tree- and distance-based metrics of phylogenetic beta diversity he explored are highly correlated with one of the pairwise or nearest neighbor metrics to be discussed below, indicating that many of the metrics that have been published are largely redundant. As such we will focus one general type of pairwise and one general type of nearest neighbor metric.

5.4.1   Pairwise Measures

In many cases we are simply interested in ecology in the overall similarity or dissimilarity between two samples or communities. In a distance-based framework this overall similarity can be calculated as the distance between all species or individuals in one community to all species or individuals in a second community. If there is no information regarding the evolutionary history or function of the species or individuals, then this metric simply calculates the overall overlap in the identities of species or individuals. The lack of information regarding the evolutionary history or function of species or individuals reduces the information we can glean from measures of community dissimilarity and even in some cases can cause us to overturn the inferences we made from information regarding identity (i.e., species names) alone (e.g., [109, 110]).

The calculation of a distance-based metric to quantify the overall dissimilarity between two samples has its roots in Rao [68], which we will discuss above. The metrics utilized today are generally natural and minor extensions to the original Rao metric. For example, the pairwise phylogenetic or trait distance (D_{pw}) between two communities can be calculated as

$$D_{pw} = \frac{\sum_i^{nk_1} \sum_j^{nk_2} \delta_{ij}}{nk_1 \times nk_2}$$

where there are nk_1 species in community k_1, nk_2 species in community k_2, and δ_{ij} is the phylogenetic or functional distance between species i in community k_1 to all species in community k_2. In this instance species i or species j can occur in both communities and conspecific distances (i.e., zeros) are counted in the mean pairwise distances between communities.

The calculation of D_{pw} in R for phylogenetic or trait data requires a phylogenetic distance matrix with all the species in the first community as rows and all species in the second community as columns. A mean of this distance matrix, including any zeros, is the D_{pw} value. To accomplish this we must first make a distance matrix for our phylogeny or trait dataset. The phylogenetic distance matrix can be generated using the `cophenetic()` function.

```
> p.dist.mat <- cophenetic(my.phylo)
```

A trait distance matrix can be calculated from the raw trait values using the `dist()` function or by using `cophenetic()` on a trait dendrogram. In this example, we will calculate a Euclidean distance matrix from the raw trait values.

```
> traits <- read.table("beta.exmample.traits.txt", sep
= "\t", row.names = 1, header = T)

> f.dist.mat = as.matrix(dist(traits, method =
"euclidean"))
```

The next step is to extract the names of the species present in our community. Here we will utilize the first community in our community data matrix.

```
> com.spp.1 <- names(my.sample[1, my.sample[1, ] > 0])

> com.spp.1
```

The result is a vector of names for those species present in the first community. This can be repeated for the second community in our community data matrix.

```
> com.spp.2 <- names(my.sample[2, my.sample[2, ] > 0])
```

Now that we have vectors containing the names of the species present in our two communities to be compared we can select the rows in our phylogenetic distance matrix corresponding to the species present in the first community and the columns corresponding to the species present in the second community. Here we are just using the phylogenetic distance matrix, but the exact same procedure could be done using the trait distance matrix to measure functional beta diversity.

```
> p.dist.mat[com.spp.1, com.spp.2]
```

An average of the resulting phylogenetic (or functional) distance matrix comparing the two communities can now be quantified to provide the D_{pw} value.

```
> mean(p.dist.mat[com.spp.1, com.spp.2])
```

As we discussed above the D_{pw} metric quantifies a mean phylogenetic or functional distance between two communities including the distance between conspecifics. If the use of conspecifics was undesirable for some reason they could easily be removed from the analysis by replacing all zero values in the community phylogenetic or functional distance matrix with NA values. In particular, we first generate a new distance matrix identical to the original.

```
> new.p.dist.mat <- p.dist.mat
```

Second we replace all instances of zero (i.e., all congeneric comparisons) with a NA. Note that in the event that your phylogeny has a branch that ends with two species or more that are not separated by any branch length or you have two or more species that have the exact same trait values the code below may present unwanted results.

```
> new.p.dist.mat[new.p.dist.mat == 0] = NA
```

We can now take a mean of the resulting community phylogenetic (or functional) distance matrix assuring that we remove all NA values.

```
> mean(p.dist.mat[com.spp.1, com.spp.2], na.rm = T)
```

Next we would like to extend the D_{pw} metric to include the abundances of species. It is critical that the abundance-weighted metric is a natural mathematical extension of the original unweighted metric and, as we have seen above, this is not always accomplished. With this in mind, we can define an abundance-weighted version of the pairwise dissimilarity (D_{pw}').

$$D_{pw}' = \sum_{i}^{nk_1}\sum_{j}^{nk_2}\delta_{ij}f_if_j$$

where f_i is the abundance of species i in community one and f_j is the abundance of species j in community two. We could alternatively write the equation with the summations in the numerator going across all individuals instead of species and the with no abundance parameters. Now that we have defined an abundance-weighted pairwise dissimilarity metric that is a natural extension of the unweighted metric, we can proceed to calculate it in R using an example quantifying the phylogenetic dissimilarity between two communities. We start by transforming our community data matrix that contains the raw count of individuals into relative abundances by dividing the values in each row by the sum of the values in that row.

```
> my.ra.sample <- my.sample / rowSums(my.sample)
```

Next, we obtain the names of the species present in the first two communities.

```
> com.1.ra <- my.ra.sample[1, my.ra.sample[1, ] > 0]

> com.2.ra <- my.ra.sample[2, my.ra.sample[2, ] > 0]
```

We then generate a community phylogenetic distance matrix containing only the species present in our two communities.

```
> new.p.dist.mat <- p.dist.mat[com.1.ra, com.2.ra]
```

Last, we calculate the abundance-weighted pairwise phylogenetic dissimilarity (D_{pw}') by calculating the product of community phylogenetic distance with the product of all abundances between species in the two communities. This value is then summed to provide the dissimilarity.

```
> Dpw.prime <- sum(new.p.dist.mat * outer(com.1.ra,
com.2.ra))

> Dpw.prime
```

The values for D_{pw} and D_{pw}' for all pairs of communities in your community data matrix can be calculated by using the comdist() function in the *picante* package, but it is important to note this function will not remove conspecifics from the analysis. Whether you calculate D_{pw} or D_{pw}' using comdist() depends on whether you tell the function true or false regarding abundance weighting.

```
> comdist(my.sample, cophenetic(my.phylo),
abundance.weighted = F)
```

When working with large datasets and/or when you are interested in running a null model analysis (see Chap. 6) you likely may find that the `comdist()` function is too slow to accomplish your tasks in a timely manner. This is because a difficulty arises when coding the calculation of beta diversity or dissimilarity metrics that is not present when calculating alpha diversity. That problem is that all pairwise comparisons of communities must be calculated. Thus, the number of calculations and outputs is much larger for beta diversity than alpha diversity. As we have seen in the previous chapters alpha diversity is often calculated for each community by writing a `for()` loop to analyze one community (i.e., one row in the community data matrix) at a time. The use of loops can severely slow down analyses and we have overcome those limitations for alpha diversity by using the `apply()` function instead of a loop. The problem is even greater for beta diversity because functions such as `comdist()` use nested `for()` loops. In other words, there will be a loop starting with community one and a loop inside that loop to compare community 1 to all of the other communities. The effect of these nested loops on beta diversity calculations may be minimal on small datasets, but on larger datasets or when you would like to analyze hundreds or thousands of random communities in a null modeling context the effect can be very large. Thus, we should attempt to find a way to utilize apply() functions to eliminate loops as much as possible in phylogenetic or functional beta diversity analyses.

Here I present one such approach for calculating the D_{pw} value for all pairs of communities in our dataset without using a `for()` loop. This will require that we write three small functions. We start by generating a simple function we will call `get.presents()` that simply reports the names of the species present in a community (i.e., a row in our community data matrix).

```
> get.presents <- function(x){

    names(x[x > 0])

}
```

We can now apply our get.presents() function simultaneously to all rows of our community data matrix to produce a list containing the names for the species present in each of our communities.

```
> list.of.names <- apply(my.sample , 1, get.presents)

> list.of.names
```

Note it is critical for the following steps that the `list.of.names` object is truly a list and not a matrix. It would only be a matrix if all communities in your community data matrix had the same exact species richness. If your object is a matrix you would use an `apply()` function in the following code with a MARGIN = 2 instead of a `lapply()` function. The next goal is to write a function to calculate D_{pw} that can be applied to each group of names in our list of species present in each community.

```
> Dpw.apply.function <- function(x){

        ## Write a function within our main function.
        ## This function calculates the mean phylogenetic
        ## distance of all species found in community z
        ## and community x. The community x comes from
        ## our main function and the community z comes
        ## from this function. This is, in essence, how
        ## we replace two nested loops.
        tmp.function <- function(z){

             mean(p.dist.mat[x, z])
        }

        ## Apply our tmp.function from above to all
        ## levels in our list.of.names list. In other
        ## words, for community x, we will calculate the
        ## Dpw between that community and all other
        ## communities including itself.
        lapply(list.of.names, FUN = tmp.function )


    }
```

The above functions will now generate a list of names present in each community, and then for a single community calculate the D_{pw} between it and all other communities. Thus to finish the analysis we have to compare all communities to all others. To accomplish this we can use an `lapply()` function. Thus we have one `lapply()` nested within the `Dpw.apply.function` and one lapply wrapped around that function. This replaces the use of two `for()` loops and vastly speeds up the calculation.

```
> dpw.output <- lapply(list.of.names,
Dpw.apply.function)
```

The result is a list with one layer for each community. Inside each layer is a list of D_{pw} values comparing that community to the others. The goal now is to put these levels back together to form an output matrix. This can be done simply using the `do.call()` function.

```
> do.call(cbind, dpw.output)
```

As similar approach can be used to calculate the D_{pw}' where the `tmp.function()` is modified to include the calculation for the D_{pw}' between two communities. We will now transition to the calculation of nearest neighbor distance-based phylogenetic and functional beta diversity metrics.

5.4.2 Nearest Neighbor Measures

The previous subsection dealt with calculation of pairwise distance-based phylogenetic and functional beta diversity calculations. While pairwise metrics maybe interesting for comparing the overall dissimilarity between sets of communities, it might be equally interesting to quantify whether the closest relative of each species or individual in one community to the next is distantly or closely related. This is the question that nearest neighbor distance-based beta diversity analyses ask and they are conceptually similar to beta diversity indices that simply ask how many species or genera are shared between two communities. A nearest neighbor extension of the alpha diversity metric *mntd* to beta diversity (D_{nn}) can be formulated as follows:

$$D_{nn} = \frac{\sum_i^{nk_1} \min \delta_{ik_2} + \sum_j^{nk_2} \min \delta_{jk_1}}{nk_1 + nk_2}$$

where $\min \delta_{ik_2}$ is the minimum phylogenetic distance between species i in community k_1 and all species in community k_2, $\min \delta_{jk_1}$ is the minimum phylogenetic distance between species j in community k_2 and all species in community k_1, and n is the number of species in the respective communities. This metric has been utilized in phylogenetic and functional beta diversity analyses and provides a refined or "terminal" measure of dissimilarity [64, 80, 120, 121]. As with the D_{pw} calculations above, conspecifics are included in this D_{nn} equation such that if two communities have the identical species composition the D_{nn} value will be zero.

We will begin with a calculation of the D_{nn} metric by comparing the first two communities in our community data matrix. The goal is to first generate a phylogenetic (or functional) distance matrix containing only the species found in the community.

```
> com.spp.1 <- names(my.sample[1, my.sample[1, ] > 0])

> com.spp.2 <- names(my.sample[2, my.sample[2, ] > 0])

> p.dist.mat[com.spp.1, com.spp.2]
```

If we look at the output distance matrix we visualize that if we were able to select the minimum value in each row we would have a series of the nearest neighbor values across species in community one and if we did the same across columns we would have the nearest neighbor values across all species in community two. This can be accomplished using the `apply()` function. Using just community one as an example we will use an `apply()` function in the rows of the community phylogenetic distance matrix to extract the minimum value in each row.

```
> apply(p.dist.mat[com.spp.1, com.spp.2], MARGIN = 1,
min, na.rm = T)
```

Because the ultimate goal is to calculate the nearest neighbor values for each species in both communities we can use two `apply()` functions, one for the rows and one for the columns, and calculate a mean value to produce the D_{nn} result.

```
> mean(c(apply(p.dist.mat[com.spp.1, com.spp.2],
MARGIN = 1, min, na.rm = T),
apply(p.dist.mat[com.spp.1, com.spp.2], MARGIN = 2,
min, na.rm = T) ))
```

In many instances it might be interesting to remove conspecific species shared between the two communities for the calculation. This may assist you in determining how much the phylogenetic or functional dissimilarity that you have found is simply the result of little species turnover or the turnover of species that are very closely related or functional very similar. We can accomplish this by first replacing all zeros in the distance matrix with NA values. Again note that this will assign NA values to non-conspecifics that have no phylogenetic branch length separating them or that share identical trait values and as such you will want to take care before applying this approach.

```
> p.dist.mat.noc <- p.dist.mat

> p.dist.mat.noc[p.dist.mat.noc == 0] = NA

> mean(c(apply(p.dist.mat.noc[com.spp.1, com.spp.2],
MARGIN = 1, min, na.rm = T),
apply(p.dist.mat.noc[com.spp.1, com.spp.2], MARGIN =
2,min, na.rm = T) ))
```

The D_{nn} value can be calculated for all pairs of communities using the comdistnt() function in the *picante* package, which has the options to weight all species equally and to exclude conspecific species. It should be noted that the comdistnt() function also simply replaces zero values in the distance matrix with NA values and therefore may place NA values for non-conspecific pairs in the distance matrix. This function also uses nested for() loops for its calculation and therefore may be much slower than the apply() function-based code above particularly with large datasets.

```
> comdistnt(my.sample, cophenetic(my.phylo),
abundance.weighted = F, exclude.conspecifics = F)
```

Like all of the other phylogenetic and functional diversity metrics presented in this book it is often very useful to weight the metrics by abundance. We will call the abundance-weighted version of the nearest neighbor metric D_{nn}'. The D_{nn}' equation for simply weights all of the nearest neighbor distances by the abundance of the focal species and averages across all individuals.

$$D_{nn}' = \frac{\sum_{i}^{nk_1} \min \delta_{ik_2} \times f_i + \sum_{j}^{nk_2} \min \delta_{jk_1} \times f_j}{\sum_{i}^{nk_1} f_i \times \sum_{j}^{nk_2} f_j}$$

where f_i is the abundance of species i in community k_1 and f_j is the abundance of species j in community k_2. The original D_{nn} calculation and the present calculation include conspecific species, but this can be easily modified to remove conspecifics. The calculation of this metric in R first requires that we convert our community data matrix to relative abundances by dividing values in the rows by the sum of the values in that row.

```
> my.ra.sample <- my.sample/rowSums(my.sample)
```

We then derive a phylogenetic distance matrix that contains all the species present in community one in the rows and all species present in community two in the columns.

```
> new.dist.mat <- p.dist.mat[names(my.ra.sample[1,
my.ra.sample[1, ] > 0]),   names(my.ra.sample[2,
my.ra.sample[2, ] > 0])]
```

The calculation is now the same as before with the exception that we calculate a weighted mean with the abundances of the species in the two communities serving as the weights.

```
> weighted.mean(c(apply(new.dist.mat, MARGIN = 1,min,
na.rm = T),apply(new.dist.mat, MARGIN = 2,min, na.rm =
T)), c(my.ra.sample[1, my.ra.sample[1, ] > 0],
my.ra.sample[2, my.ra.sample[2,] > 0]))
```

The code above can again be modified to remove conspecific species from the calculation by replacing all zeros in the phylogenetic (or functional) distance matrix with NA values. The D_{nn}' values for all pairs of communities can be calculated in the *picante* package using the comdistnt() function.

```
> comdistnt(my.sample, cophenetic(my.phylo),
abundance.weighted = T, exclude.conspecifics = F)
```

Again, like the comdist() function, the comdistnt() function has nested for() loops that can slow computation of values. Here I will quickly provide an example for how to eliminate these nested loops using apply() and lapply() functions. We begin by using the same functions to extract the names of species present in each community in our community data matrix.

```
> get.presents <- function(x){

    names(x[x > 0])

}

> list.of.names <- apply(my.sample , 1, get.presents)
```

Next we write the function that will quantify the D_{nn} between one community and all other communities.

```
> Dnn.apply.function <- function(x){

    ## Here is the only change from the code in the
    ## previous subsection for Dpw.
    tmp.function <- function(z){
    mean(c(apply(p.dist.mat[x, z], MARGIN = 1, min,
    na.rm = T), apply(p.dist.mat[x, z], MARGIN = 2,
    min, na.rm = T) ))
    }

lapply(list.of.names, FUN = tmp.function )

}
```

Now that the function has been written to calculate the D_{nn} between one community and all others, we can use `lapply()` to apply this function to all levels in our list containing the names of species present in each of our communities. Recall that the output of the `lapply()` last time was a list of values for each community. We will therefore wrap our code with `do.call()` to zip our output back together into a matrix that is easy to visualize.

```
> do.call(cbind, lapply(list.of.names,
Dnn.apply.function))
```

A similar procedure can be used to calculate the abundance-weighted nearest neighbor metric by replacing the code in the `tmp.function()` with a weighted mean calculation.

5.5   Other Metrics

The above sections cover the major tree- and distance-based approaches and metrics for calculating phylogenetic and functional beta diversity. Like any other series of diversity metrics, novel and not-so-novel metrics probably will be published while this book is being written and after. We cannot cover those metrics, of course, and we won't cover many other metrics that are monotonic with the distance-based metrics provided above with one exception. We will briefly discuss the dissimilarity indices of Rao [68]. We discuss these indices given that they were published long before the metrics above and therefore have a special place in phylogenetic and functional diversity metric history even though their application to this type of data was not immediate.

The two Rao dissimilarity metrics that we will briefly discuss can be calculated with the `raoD()` function in the *picante* package using a community data matrix and a phylogeny. This function does not take a trait distance matrix, but given the results are monotonic with the above metrics we will not worry about this issue since one can simply use those metrics instead.

```
> rao.output <- raoD(my.sample, my.phylo)
```

The first dissimilarity metric output by the `raoD()` function is what I will call Rao's D_{beta}. The equation for this metric is defined simply as the pairwise phylogenetic (or functional) distance between all individuals in two communities.

$$Rao\text{'s}\, D_{beta} = \sum_i \sum_j \delta_{ij} f_{ik_2} f_{jk_1}$$

Thus, we can see that this metric is essentially the same as the D_{pw}' metric, but in this instance the δ_{ij} value is equal to the distance to the most recent common ancestor (MRCA) of two species and not the complete distance between two species. In other words, D_{pw}' will simply be equal to twice the Rao's D_{beta} for an ultrametric phylogeny. The Rao's D_{beta} can be extracted from the output we generated above by asking for the Dkl matrix.

```
> rao.output$Dkl
```

Rao has also proposed a metric, which I will call Rao's H, that weighted the Rao's D_{beta} by the diversity within each community.

$$Rao\text{'s}\, H = \frac{Rao\text{'s}\, D_{beta}}{\left(\sum_i^{S_{k_1}} f_i \overline{\delta_{k_1}} + \sum_j^{S_{k_2}} f_j \overline{\delta_{k_2}} \right) \times \frac{1}{2}}$$

Again it is essential to recall that the phylogenetic distance values in this equation are distances to the MRCA and not the complete distance separating two taxa on the phylogeny. The values for Rao's H can be extracted from the output we produced by asking for the H matrix.

```
> rao.output$H
```

Now that we have covered the original Rao metrics we can transition into comparing all of the phylogenetic and functional beta diversity metrics covered in this chapter.

5.6 Comparing Metrics

As we have discussed in the two previous chapters regarding alpha diversity it is often useful and informative to compare metrics. This helps us to understand which metrics are monotonic and which are providing unrelated information that

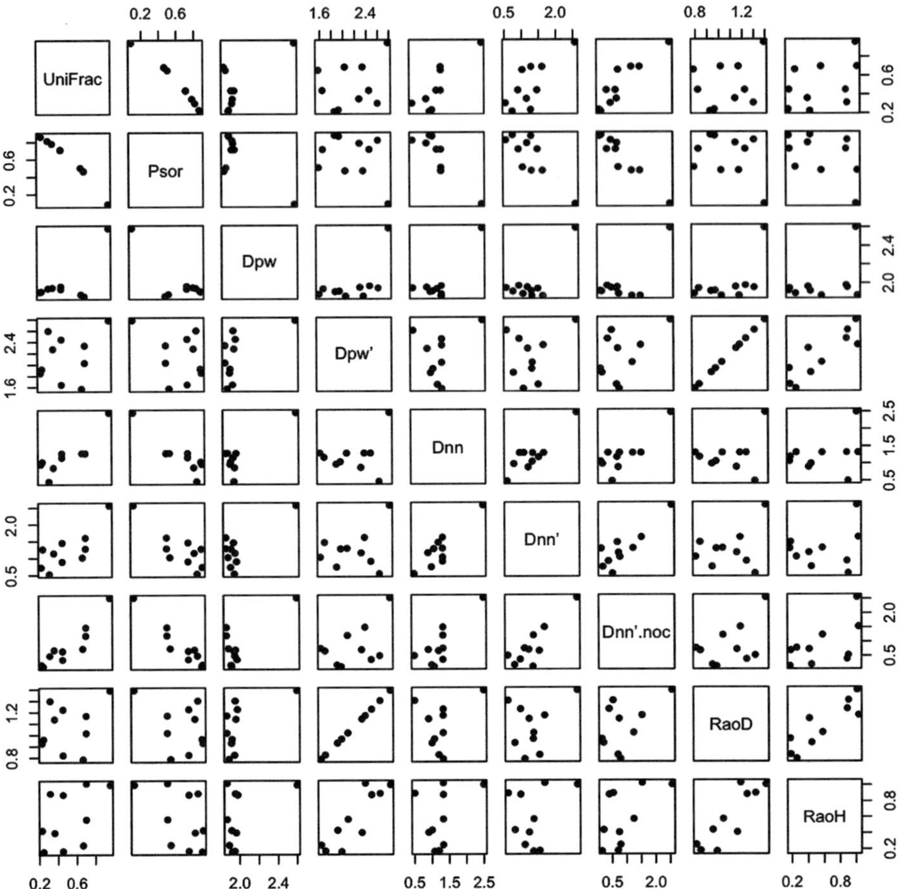

Fig. 5.2 A plot of each phylogenetic beta diversity metric against all other metrics. Note that while some metrics are not related others are highly correlated in this dataset. For example the D_{pw}' and Rao D metrics are all highly correlated and are in fact redundant

may be helpful. This is important for our understanding of the metrics published and outlined here, but also if you are considering coding a new metric for your research. If your "new" metric is monotonic with existing metrics, then it probably is not needed and shouldn't be published in order to avoid cluttering and in some cases mathematical synonymy in the already vast diversity metrics literature (Fig. 5.2).

```
> unifrac.output <- c(unifrac(my.sample, my.phylo))

> p.sor.output <- c(phylosor(my.sample, my.phylo))

> Dpw <- c(comdist(my.sample, cophenetic(my.phylo),
abundance.weighted = F))

> Dpw.prime <- c(comdist(my.sample,
cophenetic(my.phylo), abundance.weighted = T))

> Dnn <- c(comdistnt(my.sample, cophenetic(my.phylo),
abundance.weighted = F, exclude.conspecifics = T))

> Dnn.prime <- c(comdistnt(my.sample,
cophenetic(my.phylo), abundance.weighted = T,
exclude.conspecifics = T))

> Dnn.prime.noc <- c(comdistnt(my.sample,
cophenetic(my.phylo), abundance.weighted = T,
exclude.conspecifics = F))

> rao.D <- c(as.dist(raoD(my.sample, my.phylo)$Dkl))

> rao.H <- c(as.dist(raoD(my.sample, my.phylo)$H))

> outputs <- as.data.frame(cbind( c(unifrac.output),
c(p.sor.output), c(Dpw), c(Dpw.prime), c(Dnn),
c(Dnn.prime), c(Dnn.prime.noc), c(rao.D), c(rao.H)  ))

> names(outputs) <- c("UniFrac", "Psor", "Dpw", "Dpw'",
"Dnn", "Dnn'", "Dnn'.noc", "RaoD", "RaoH")

> plot(outputs, pch = 16)
```

We can see that metrics such as Rao's D and D_{pw}' and UniFrac and PhyloSor are monotonic and highly correlated if not mathematically identical. We can test degree of correlation with a simple Pearson's correlation.

```
> cor(outputs)
```

Thus, we can now quickly see which metrics are clearly monotonic or at least highly correlated. The results are to some degree particular to the datasets we are using, but the general direction and strength of the correlations will likely be seen in almost any dataset you use. For example, the perfect correlation between Rao's D and D_{pw}' is a mathematical necessity and it is unlikely, given how they are calculated, that UniFrac and PhyloSor will ever be divergent across many communities or samples in a system. This underscores the importance of knowing why you are calculating one metric over another or both metrics and knowing whether we should expect them to tell us novel information.

5.7 Conclusions

Here we have focused on phylogenetic and functional measures of beta diversity. There is approximately a decade worth of research regarding phylogenetic beta diversity with the vast majority of the early work coming from microbial ecologists that necessarily had to use phylogenetic information to characterize the beta diversity in their systems. Functional beta diversity research is much rarer, but is becoming more prevalent in the literature over the past few years. The promise of both approaches is that by recognizing that species are not independent entities and by quantifying their phylogenetic and functional non-independence we can provide stronger tests of ecological and evolutionary hypotheses. These tests can be easily facilitated in R using the code provided in the above sections. In particular, we have discussed the two main classes of phylogenetic and functional beta diversity metrics that have been generated and a central framework for quantifying a metric from each class that is flexible. We have also discussed other tree-based metrics of phylogenetic beta diversity that are more common in the microbial ecology literature that rely on quantifying the shared branch length between communities and less on the phylogenetic distance between species or individuals in two communities. These tree-based metrics are, for the most part, less flexible and computationally slower and therefore are less often used by plant and animal ecologists. It is inevitable that more phylogenetic or functional beta diversity metrics will be published in the near future or there are metrics that already exist that I have not covered. It will be a challenge to navigate this barrage of new metrics, but using the code in this chapter you should be able to quantify these new metrics in R and compare them to existing metrics to determine their novelty and when, where, and why they may be useful in your research program.

5.8 Exercises

1. Quantify the species beta diversity in your example dataset using the presence–absence-weighted Jaccard's Index and the abundance-weighted Bray–Curtis Index. Use the `vegdist()` function in the *vegan* package for both metrics.
2. Correlate both measures of species beta diversity with the phylogenetic and functional beta diversity metrics presented in this chapter. Are the phylogenetic and functional metrics correlated with the species beta diversity metrics? Is this important?
3. Simulate a coalescent phylogeny in R that has the same number of species in your example dataset using the `rcoal()` function in *ape*. Take the species names from your community data matrix and place them on the tips of your simulated phylogeny.
4. Calculate the phylogenetic beta diversity metrics addressed in this chapter using the simulated phylogenetic tree and your original community data matrix.

Chapter 6
Null Models

6.1 Objectives

The ultimate goal of this chapter is to understand when, where, and why null models should be used in the analysis of phylogenetic and functional diversity. The specific objectives of this chapter are to first discuss the philosophy behind null models, what they seek to accomplish, and how they work. Second is to establish why null models are necessary for most analyses of phylogenetic and functional diversity. Lastly, we will implement several classes of null models for phylogenetic and functional alpha and beta diversity.

6.2 Background

The usage of null models in ecology has a history dating back, to the best of my knowledge, nearly 100 years [52]. Interestingly, the first null models were used in the context of relatedness and co-occurrence where researchers were quantifying the ratio between the number of genera per community to the number of species per community. A lower ratio indicating a lower degree of relatedness and higher phylogenetic diversity—although it was not termed phylogenetic diversity at this stage. A higher ratio indicated a higher degree of relatedness and less phylogenetic diversity. Not long after the first studies of these ratios were published, the question of what ratio might one expect given the observed richness was raised. In other words, it was not possible to tell whether the pattern observed was unusual or could be explained simply by random colonization of the community from a given pool of species. This fundamental problem spurred the use of null models in ecology generally and community ecology in particular [52].

The online version of this chapter (doi: 10.1007/978-1-4614-9542-0_6) contains supplementary material, which is available to authorized users

N.G. Swenson, *Functional and Phylogenetic Ecology in R*, Use R!,
DOI 10.1007/978-1-4614-9542-0_6, © Springer Science+Business Media New York 2014

In the early null modeling literature there was typically not enough computing power available to generate random communities from a given source pool that could be used to generate a null distribution to which the observed could be compared. This resulted in some analytical solutions and in some cases randomizations done by hand. In these studies and those that were the first performed on computers, generally the expected co-occurrence of species or their relatedness based on a genus:species ratio was generated by randomly assembling communities from a source pool of names while maintaining the observed number of species. In other words, if your community had three species observed, you would randomly choose three species from a source pool of names; calculate the statistic of interest (e.g., genus:species ratio) and repeat this process hundreds of times to generate a null distribution. With this approach the only pattern observed that is fixed in the randomizations is the observed species richness. Thus the null models were fairly "unconstrained."

The modern conceptual approach to null modeling is to attempt to fix all observed patterns in the data *except* the one pattern that you are interested in studying [122]. Thus, by only fixing the observed species richness of a community, a number of other observed patterns (e.g., species rarity) were free to vary. This mismatch between concept and implementation generally resulted in high type I error rates, but this mismatch and the increase in computational power also spurred methodological advancements that permitted researchers to increasingly constrain their null models such that fewer observed patterns varied in the randomizations. Though, in some cases early on and still today null models may be so constrained that there is no power to detect anything [123, 124]. Thus, when conducting null modeling analyses it is critical to: (*1*) obey the first rule of null modeling which is to fix all observed patterns in the randomizations except the pattern of interest and (*2*) be cognizant of the inherent interactions between type I error rates, statistical power, and the degree of constraint imposed in your null model. In the following sections we will describe exactly why null models are useful for phylogenetic and functional diversity analyses, what types of null models are typically used, and their relative degree of constraint.

6.2.1   Why Use Null Models for Phylogenetic and Functional Analyses?

The calculation of phylogenetic diversity (PD) or functional diversity (FD) itself can be the end goal in many analyses. For example, one may be interested in whether the PD contained in one proposed nature reserve is greater than that of another proposed reserve or whether FD is a significantly related to an ecosystem process or service. Such approaches are legitimate, but they are also generally working under the implicit or explicit assumption that PD or FD measures are providing information that measures of species richness or diversity do not. This additional information is thought to provide refined or orthogonal information regarding, for example, the conservation decision being made or the ecological process being studied. In many cases the PD or FD metrics being utilized by biologists are highly

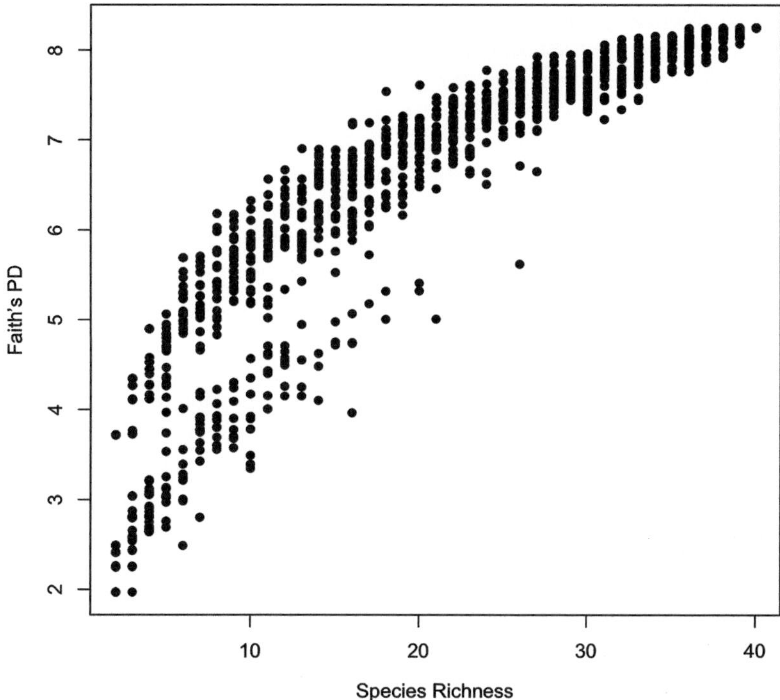

Fig. 6.1 A plot of the Faith's Index PD value from a community against the species richness from the same community. Two aspects of this plot are important. First, the two variables are highly correlated. Second, the variance decreases as one moves from species-poor to species-rich assemblages. This decrease in variance demonstrates why analyzing the residuals of this relationship alone does not remove the dependence of Faith's Index on species richness

correlated with the species richness (SR) of the community. The PD or FD measure in these instances may refine our quantification of biodiversity by adding an evolutionary or functional dimension, but the strong correlation between SR and PD leaves the question of how much additional information is actually being gained. This can be demonstrated by calculating the Faith's Index PD metric for an example dataset (Fig. 6.1).

```
> pd.sample <- readsample("null.pd.example.sample.txt")

> pd.phylo <- read.tree("null.pd.example.phylo.txt")

> faith.output <- pd(pd.sample, pd.phylo)

> plot(faith.output[,2], faith.output[,1], xlab =
"Species Richness", ylab = "Faith's PD", pch = 16)
```

The benefit of metrics for PD or FD such as Faith's Index [25] and Petchey and Gaston's analog FD [76], respectively, is that the value must increase as species are added to an assemblage. In other words, adding any species will add to the biodiversity. This makes intuitive sense, but it results in a strong correlation between SR and PD or FD when there is even a moderate gradient in SR. This creates several problems. For example, if a researcher reports that two communities have very different PD values, it can be impossible to know if this is simply because they have very different SR values or there is something fundamental about the phylogenetic information that is important. An enticing method for removing this artifact to determine whether the differences in PD or FD between assemblages are different than that expected given the SR is to examine the residuals of the SR–PD or SR–FD plot with residuals below the regression line indicating lower than expected PD or FD given the SR in a particular community and residuals above the regression indicating higher than expected PD or FD given the SR in a particular community. The problem with this particular approach, that is not always appreciated, is that the variance of the residuals around the regression must decrease as the SR in a community approaches the number of species in the phylogeny due to how tree-based metrics that are additive with SR are calculated. The magnitude of the variance with SR will also differ depending on the shape and size of the phylogenetic tree in the study thereby making comparisons of residuals impossible. Lastly, whether the observed PD or FD is significantly different from the value predicted by a regression fit cannot be determined easily and calculating a P-value using this approach is pointless. In sum, it is impossible to know whether the raw PD or FD value generated from many metrics is truly telling us something aside from the SR of the assemblage and simple calculations of how the metric deviates from the expected using a regression approach are invalid. For these reasons null models are generally advised for any analysis of PD or FD.

The knowledge that some metrics of PD or FD are strongly correlated with SR has lead researchers to develop metrics of PD or FD that are not correlated with SR. The argument made by these researchers is that the metrics are not correlated and are therefore providing orthogonal information regarding the biodiversity in an assemblage. Although this philosophy removes the appealing aspect of other metrics where the addition of a species increases the PD or FD, it does add the appealing aspect that SR and PD or FD may be treated as independent variables in an analysis. In other words, it may be possible to use SR and FD as uncorrelated independent variables with an ecosystem process as the dependent variable to determine the relative contribution of each dimension of biodiversity. The majority of the PD or FD metrics proposed that are uncorrelated with SR involve a pairwise calculation. For example, the mean pairwise phylogenetic distance between all species in a community is one such metric. We can quickly calculate this metric and demonstrate the lack of correlation with SR (Fig. 6.2).

```
> mpd.output <- mpd(pd.sample, cophenetic(pd.phylo),
abundance.weighted = F)

> plot(faith.output[,2], mpd.output, xlab = "Species
Richness", ylab = "MPD", pch = 16)
```

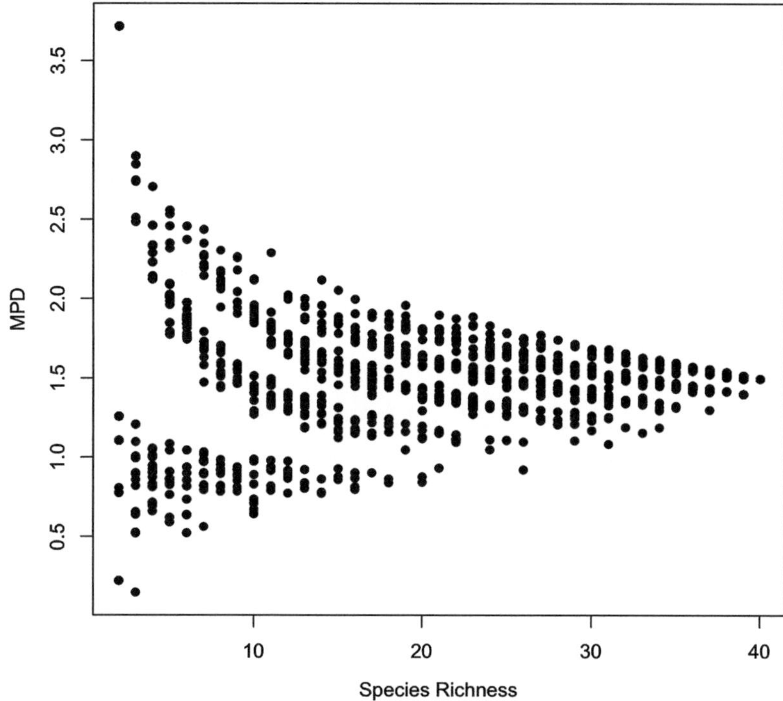

Fig. 6.2 A plot of the Mean Pairwise Phylogenetic Distance (MPD) value from a community against the species richness from the same community. One aspect of this plot is particularly important. The variance in MPD decreases as one moves from species-poor to species-rich assemblages. This decrease in variance demonstrates why analyzing the residuals of this relationship alone does not remove the dependence of MPD on species richness. It further demonstrates that although these metrics are not correlated, this does not mean that MPD itself is truly independent from species richness because a greater range of MPD values is mathematically possible at low-species richness values. Thus, researchers will tend to find "less significant" MPD values in assemblages with higher species richness if a proper null model is not utilized

We see there is no correlation between the metric and SR. Similar results would be found using other pairwise calculations such as FDis, Rao's Index, and PSV. This type of plot generally forms the basis for the argument that the raw values of these metrics can be used as independent biodiversity information for downstream analyses. Despite the vigor with which some make this argument, there is still a fundamental bias related to SR that makes comparisons using these raw values ill advised. Indeed, the very nature of how pairwise metrics are calculated is problematic and is in a sense the same exact problem with the tree-based metrics described above. That is, as the SR of the community approaches the number of species in the phylogeny or trait dataset, the range of possible PD or FD values calculated from a pairwise metric decreases. In other words, the mathematical mechanics of these metrics dictates that the variance in PD or FD has to decrease as SR increases where the values converge on the mean of the phylogenetic or functional distance matrix being utilized. Thus, the very high or very low PD or FD values that are possible with these metrics at low SR are mathematically not possible at high SR. This is one reason why plots of these PD or FD metrics against SR demonstrating no correlation often have a "funnel shape" with decreasing variance along the SR axis. In sum, even those metrics that are designed to be independent of SR and can be shown to be uncorrelated have an underlying bias associated with SR that makes the comparison of the PD or FD between communities difficult particularly when the difference in SR between the two communities is large. For these reasons null models are generally advised for *any* analysis that seeks to utilize PD or FD values for inference including those metrics that are uncorrelated with SR.

6.2.2 *Calculating Standardized Effect Sizes, Quantiles,* *and* **P-Values**

The problems associated with comparing the PD or FD values across communities and assemblages can generally be summarized as being the result of decrease in the range of possible values as the SR of a community or assemblage approaches the number of species in the phylogeny or trait dataset. To permit comparative analyses, we must remove this artifact by quantifying whether the PD or FD is higher or lower than expected given the observed SR in the community or assemblage. As we discussed above, some have utilized residual analyses of SR–PD or SR–FD plots to calculate the magnitude and direction of the deviation of the PD or FD from the expectation derived from the regression line. The necessary decrease in the range in possible values with SR means that comparing these raw deviations from the regression line, or "effect sizes," is hopelessly biased such that the absolute value of the magnitude of the deviations will always be much higher at lower SR values. Further, a regression approach is ill advised because it is often difficult to estimate the range of possible values or the variance of possible values at a given SR value using the observed dataset.

Null models permit researchers to estimate the expected distribution of PD or FD values for a given SR. This expected, or null, distribution can be used to calculate an effect size (*E.S.*) by measuring the deviation of the observed value from the mean of the null distribution.

$$E.S. = observed - \overline{null}$$

where positive *E.S.* values indicate an observed value higher than the average expected value and negative *E.S.* values indicate an observed value lower than the average expected value. Given the range of possible PD or FD values and the possible variance in values decreases with SR, not all *E.S.* values are created equal. That is, an *E.S.* value of 5.24 for one community containing 5 species is not comparable to an *E.S.* value of 5.24 for a second community containing 45 species. Indeed, the 5.24 *E.S.* value for the community with 45 species is much more impressive given the lower possible maximum value. A standardization of *E.S.* values is therefore necessary before they can be compared without bias. A standardized effect size (*S.E.S.*) is generally calculated in null model analyses by dividing the *E.S.* by the standard deviation of the null distribution.

$$\text{S.E.S.} = \frac{observed - \overline{null}}{sd\left(null\right)}$$

This calculation removes any directional bias associated with the decrease in variance in the expected values with increasing SR. Similar to the *E.S.* calculation, a positive value *S.E.S.* indicates a higher than the average expected value and a negative value indicates a lower than the average expected value. The calculation of *S.E.S.* values using output from a null model is an effective method for comparing the PD or FD of the communities and assemblages in your study system while removing biases associated with differences in SR. It is important to remember, however, that the shape of the null distributions may be variable across studies making meta-analysis difficult. For example, the null distributions in one study may be highly skewed and approximately normal in another study. Part of this bias could be removed if the *S.E.S.* values were reanalyzed using the median of the null distribution instead of the mean. This highlights the importance of examining the shape of some or all of the null distributions used in your study and it also highlights the usefulness of calculating quantile values in studies that utilize null models.

The quantile value or score is simply where the observed value lands in the null distribution. These scores can be utilized to estimate *P*-values. For example, an observed value that lands in the middle of a normal null distribution containing 999 values will have 500.5 as its quantile score. Or if an observed value is higher than any of the 999 random values, it will have a quantile score of 1,000, and if the observed is lower than any of the random values, it will have a quantile score of 1. Thus, a study hypothesizing that the PD in a community will be lower than expected with $\alpha = 0.05$ (i.e., a one-tailed test) will need a quantile score for the observed value is less than or equal to 50 to reject the null because $50/1,000 = 0.05$. If the hypothesis is that the observed PD in a community will be significantly different than expected with $\alpha = 0.05$ (i.e., a two-tailed test), then the quantile score for the observed value must be less than or equal to 25 for a lower than expected PD or greater than or equal to 975 for a higher than expected PD. The calculation of quantile scores and the associated estimated *P*-values removes some of the biases that make comparing *S.E.S.* values from multiple studies difficult, but they also remove information regarding the size of the effect itself. Thus, it is recommended that investigators report both the *S.E.S.* values and the quantile scores when possible to aid those readers interested in place the result of a particular study into a broader context or distribution of results.

6.3 Classes of Null Models in Phylogenetic and Functional Analyses of Species Assemblages?

There are two general classes of null models that you will encounter when it comes to phylogenetic and functional diversity analyses. In both classes the objective is to maintain all aspects of the observed data except the pattern of interest. Traditionally, measures of phylogenetic and functional diversity have been interested in whether co-occurring species exhibit more or less phylogenetic or functional diversity, however quantified, than expected. The word "expected" in the previous sentence is key as the structure of the null model underlies what the expectation really is in your study. In most cases, researchers are interested in whether the phylogenetic or functional diversity is more or less than expected given all other observed patterns in the community including the observed species richness, the observed occupancy rates of species (e.g., how many communities a species occupies in a given meta-community), the observed species abundances, etc. The first class of null models used to address this type of question seeks to randomize the community data matrix itself and does not alter the phylogenetic tree or the trait data matrix. This approach can be unconstrained where potentially many aspects of the observed community data matrix may vary during each iteration of the null model or the approach can be constrained where the investigator seeks to fix as many observed patterns as possible (e.g., community species richness and species occupancy rates) while allowing one pattern (e.g., species co-occurrence) to vary.

An alternative approach is to randomize the phylogenetic tree or trait data while fixing the observed community data matrix. This approach is more recent in the literature, but it has the advantage not having to deal with the many interconnected patterns in a community data matrix that can be hard to isolate in a randomization framework. This is not to say that potential biases may arise in a null modeling approach that randomizes the phylogenetic or functional trait data. We will discuss where these biases may lurk and how to constrain the phylogenetic or functional trait randomizations to mitigate such biases.

We will begin our consideration of null models for phylogenetic and functional diversity analyses with the first class of models—community data matrix randomizations. The discussion will start with the most basic, unconstrained, and computationally easy null models to provide historical context and then we will quickly transition to the more constrained models that are currently used. We will then discuss the second class of null models designed to randomize phylogenetic or functional trait data while maintaining the community data matrix.

6.4 Randomizing Community Data Matrices in R

As we have established in Sect. 6.2, null models have been present in ecology for nearly a century. We have also established that good practice when conducting a null model is to attempt to fix all observed patterns or aspects of the data *except* the aspect you are interested. In the case of phylogenetic and functional analyses of communities, one is often interested in determining whether the community pattern of interest (e.g., degree of co-occurrence, species similarity in assemblages, etc.) is nonrandom with respect to phylogeny or function. This leaves several options for constructing a null model. The first we will discuss is the randomization of community data. Randomizing community data was the original null modeling approach for questions regarding relatedness and co-occurrence and therefore has a deeper literature that has documented the advantages and disadvantages of various community data randomizations. Given that almost all analyses of phylogenetic and functional diversity in R require a community data matrix, we will focus our discussion on the randomization of these matrices while maintaining the observed phylogenetic and functional data. Recall that a community data matrix contains columns for each species in the meta-community and rows for each community in the meta-community.

6.4.1   *Unconstrained Randomizations*

The original null models utilized in ecology generated random species assemblages that generally only fixed one or a few observed patterns. Thus, these null models may be considered "unconstrained" or minimally constrained. These null models are generally quicker to compute and easier to program, but many aspects of the data are often randomized along with the pattern of interest. This typically results in an inflation of type I error and therefore they are not recommended for most analyses. Though we will discuss these null models for historical completeness and for those rare cases when unconstrained randomizations of community data matrices are appropriate.

The classic unconstrained null model is one that r-assembles the communities at random individually while maintaining the species richness observed. Say we have a meta-community containing species six species (A, B, C, D, E, and F) and in the three communities in our study system we know their presence or absences. The community data matrix would then look something like this:

```
> com.data <- matrix(c(1,0,1,0,1,0,1,1,1, 1,1,0,
1,1,1, 0,0,1), nrow = 3)

> rownames(com.data) = c("com1", "com2", "com3")

> colnames(com.data) = c("A", "B", "C", "D", "E", "F")

> com.data
```

The species richness for each community can be calculated by summing the ones in each row and we see each community has four species.

```
> rowSums(com.data)
```

If we wanted to randomly assemble community one (i.e., com1), we would then randomly draw four species from the list of A through F without replacement.

```
> sample(colnames(com.data), size = 4, replace = F)
```

This could be repeated many times to produce many null or random communities for com1. Here we will only replicate the process ten times.

```
> t(replicate(10, sample(colnames(com.data), size = 4,
replace = F)))
```

To perform this randomization for all communities to output a community data matrix we can use the `randomizeMatrix()` function in the *picante* package.

```
> library(picante)

> reps <- replicate(5, randomizeMatrix(com.data,
null.model = "richness")

> reps
```

The output is an array with each level being a random community data matrix. These random matrices could be used to calculate random phylogenetic and functional alpha and beta diversity. For example, one couple replicate the function 999 times instead of 3 and produce 999 random community data matrices and then use a function like `PD()` to calculate the Faith's Index for the 999 random community matrices.

If we take a look at the output from the randomization above generating three community data matrices we will see that while the row sums do not change (i.e., the community species richness is fixed), the column sums are not consistent for species indicating that the occupancy rate, the number of communities a species is in, is not maintained. Though there is a chance this didn't happen with your data due to the low number of randomizations we used.

```
> apply(reps, MARGIN = 3, rowSums)

> apply(reps, MARGIN = 3, colSums)
```

This variability in the column sums indicates that not only the co-occurrence of species pairs is varying, but their overall occupancy rates are also changing. Indeed a species could be present in only one community in one randomization (i.e., rare) and present in all communities in the next (i.e., common). This suggests that multiple properties of the matrix are not fixed in randomizations and this likely could inflate type I error.

6.4.2   Constrained Randomizations

A solution to the unconstrained randomization problem from the previous subsection is to fix the row and column sums for a presence–absence community data matrix in all randomizations. This approach is now known as an "Independent Swap" null model [125]. An independent swap is computationally more difficult and initially restricted researchers from implementing this approach, but it is now easily implemented in R using the same function as above:

```
> reps.is <- replicate(5, randomizeMatrix(com.data,
null.model = "independentswap")
```

We can now see that the row and column sums (i.e., the community species richness and species occupancy rate) are now consistent across randomizations.

```
> apply(reps.is, MARGIN = 3, rowSums)

> apply(reps.is, MARGIN = 3, colSums)
```

Thus, a useful and computationally expedient constrained null model is available in R for randomizing our community data matrices that contain presence and absence of species. The next goal, obviously, is to randomize a community data matrix that contains abundances or relative abundances. To explore this possibility, let us assign abundances to our species.

```
> com.data.a <- matrix(c(3,0,2,0,10,0,4,2,3, 4,5,0,
6,1,2, 0,0,7), nrow = 3)

> rownames(com.data.a) <- c("com1", "com2", "com3")

> colnames(com.data.a) <- c("A", "B", "C", "D", "E", "F")

> com.data.a
```

Now run the independent swab null model and sum the rows and columns.

```
> reps.a.is <- replicate(5, randomizeMatrix(com.data.a,
null.model = "independentswap"))

> apply(reps.a.is, MARGIN = 3, rowSums)

> apply(reps.a.is, MARGIN = 3, colSums)
```

We see that the row sums are in some instances different indicating that the total abundance of a community varies from randomization to randomization. The column sums on the other hand are consistent indicating the total abundance of a species in the system is fixed in the randomizations. Now let us examine the actual output community data matrices.

```
> reps.a.is
```

We see the occupancy rate of species (i.e., the number of row a species has a positive abundance) is fixed. That is good news. Though, unfortunately, when we consider all of this evidence we see that adding abundance to this problem that the total abundance in a community is no longer fixed. This likely will have consequences for not only abundance-weighted alpha diversity measures but also abundance-weighted beta diversity measures. Thus, if we are using abundance data it may be very difficult to fix all the observed properties of the communities aside from the one property we are interested in analyzing. This is a fundamental, and thus far unavoidable, problem for null models randomizing community data matrices. In the following sections we will consider randomizations that fix the entire community data matrix while randomizing the phylogenetic or functional information.

6.5 Randomizing Phylogenetic Data

Null models that focus on the randomization of community data matrices are becoming increasingly sophisticate and fast, but even the most constrained of these randomizations are often fairly unconstrained from my perspective. In many instances this can make it difficult to determine whether the statistical signal being produced can be unbiased and only informing the investigator about the pattern of interest. An alternative approach for phylogenetic diversity investigations, that may be less susceptible to the biases that lurk in community data matrix randomizations, is to randomize the phylogenetic relatedness of species. Thus, when asking a question such as does my community have a higher phylogenetic diversity than expected, we do not construct random communities. Rather we randomize the phylogenetic distances between the species in our community of interest. The randomization of phylogenetic relatedness therefore preserves all obvious and less obvious patterns and information in the community data matrix. Also, as you will see below, randomizations of phylogenetic relatedness are simple and computationally fast. Though, similar to community data matrix randomizations, phylogenetic randomizations can be unconstrained or constrained and care must be taken when utilizing one over the other.

6.5.1 Unconstrained Randomizations

The overarching goal of randomizing phylogenetic data for ecological analyses is generally to randomize the relatedness of species or individuals. The ultimate unconstrained phylogenetic randomization would be to simulate phylogenetic trees that contain the same number of species in your data. In the context of most ecological investigations, that would mean simulating phylogenies with the same number of taxa as that found in the observed meta-community. This is ill advised for several reasons as it not only randomizes the overall ranking of relatedness of taxa (i.e., the topology of the phylogeny) but also the distribution of relatedness of between taxa (i.e., the distribution of branch lengths). This total lack of constraint would increase the type I error rate to such a level that it is not worth demonstrating how such a null model could be constructed in R. Rather we consider a more constrained null model that randomizes the names of taxa on the phylogeny. Thus, it randomizes who is

Fig. 6.3 A plot of our example phylogeny

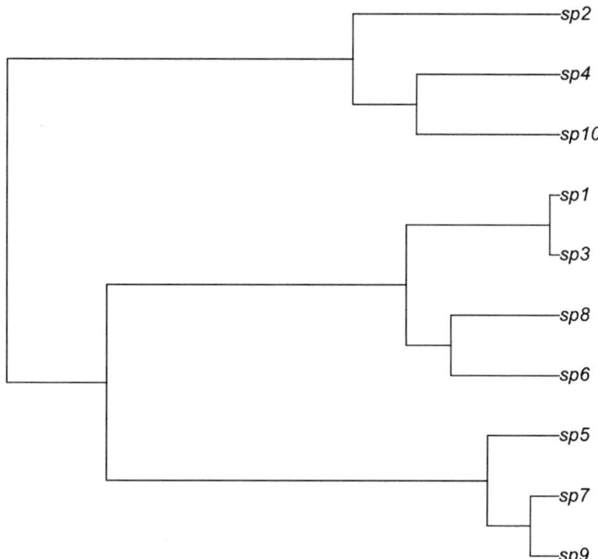

most closely related to whom, but it does not alter the actual branch lengths and their distribution.

We will begin by first reading our example Newick formatted phylogeny into R using `read.tree()`. As always we will also plot the phylogeny (Fig. 6.3).

```
> my.phylo <- read.tree("null.example.phylo.txt")

> plot(my.phylo)
```

The goal of the unconstrained phylogenetic randomization is to simply randomize the taxa names on the tips or terminal branches on the phylogeny. To accomplish this we must first consider how the taxa names are stored in our `my.phylo` object.

```
> my.phylo$tip.label
```

We see that we can obtain a vector listing the names of all taxa in the phylogeny. We see that the order of the names is the same order of the names in the plotted phylogeny from the bottom to the top of the plot the image if you use `plot(my.phylo)`. The goal therefore is to randomize the position of the names in this vector and to place them back onto the phylogeny. First, let us randomize the names in the vector using `sample()`.

```
> sample(my.phylo$tip.label,
length(my.phylo$tip.label), replace = F)
```

The resulting vector of names is in a different order than the original vector. This was accomplished by telling `sample()` to take the vector of names and to resample them without replacement until we have the same number of names as in the original vector. Because we have not permitted replacement, this approach simply shuffles the position of the names in the vector. If we permitted replacement, some names would appear twice in the new vector and some species would be omitted. This is not desired as we do not want the same taxa names appearing on multiple terminal branches in our randomized phylogenies. Now that we know how to randomize the taxa names, we must put them back onto our phylogeny. Because we do not wish to alter our original phylogeny, first make a new object that is the same as our original phylogeny.

```
> my.phylo.rand <- my.phylo
```

Next we want to place the taxa names (i.e., tip labels) on our new phylogeny object with the randomized list of names.

```
> my.phylo.rand$tip.label <-
sample(my.phylo$tip.label, length(my.phylo$tip.label),
replace = F)
```

The above code could be used to randomize the names on the original phylogeny, `my.phylo`, without having to create a separate phylogeny object, but this is not advised. To assure that the randomization of taxa names was successful, plot the randomized phylogeny and compare it to the original to assure that it is indeed randomized. First open a new quartz window where we can plot the randomized phylogeny.

```
> quartz()
```

Now plot the randomized phylogeny for comparison. Hopefully the names of the taxa on the randomized phylogeny are in a different order (i.e., on different terminal branches) than the original phylogeny (Fig. 6.4).

```
> plot(my.phylo.rand)
```

The names of the taxa should be in a different order. If they are not, run the above code again to make a second randomized phylogeny. As the size of the phylogeny decreases, the number of possible ways taxa could be arranged on the phylogeny will of course also decrease. Thus, if you are using your own phylogenetic tree and it has few taxa, then it is more likely that your resampling or shuffling of taxa across the terminal branches will result in the same distribution of names as in the original phylogeny. Now that we know how to randomize the names of taxa on the terminal branches of a phylogeny we are able to implement this procedure more broadly. Although, we can now code this randomization ourselves, let us utilize an existing function for shuffling names of taxa on a phylogeny. First load the R package *picante*, which you should already have installed:

```
> library(picante)
```

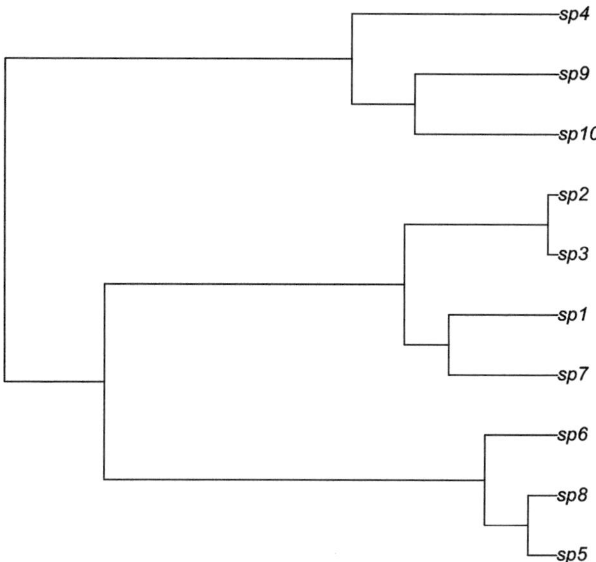

Fig. 6.4 A plot of our example phylogeny with the species names randomized. Compare with Fig. 6.3

Now use the function `tipShuffle()` to randomize the names of taxa on the phylogeny. We will assign this new phylogeny as a new object.

```
> my.phylo.rand.2 <- tipShuffle(my.phylo)
```

To assure that this function is indeed randomizing the names of taxa on the phylogeny, plot the new object and compare the distribution of names on the original and newly randomized phylogenies (Fig. 6.5).

```
> quartz()
```

```
> plot(my.phylo.rand.2)
```

We have now learned how to randomize the names of taxa on a phylogeny using our own code and the code in an existing R package. This knowledge is the fundamental building block for coding in R almost any unconstrained phylogenetic randomization for the purpose of asking whether the observed phylogenetic diversity is higher or lower than that expected. To demonstrate this we will now code our own null model for the mean pairwise distance (MPD) phylogenetic diversity metric and Faith's Index of phylogenetic diversity (PD). Although we focus on the MPD and PD metrics, the basics of the code below could apply to the use of generally any metric of phylogenetic diversity you would calculate in R.

When coding any null model in R, there are a couple of options. One option is to write `for()` loops that loop through a series of randomizations (e.g., 999). This is generally the approach most functions use despite being much slower than the second option. The second option is to use a `replicate()` function. We will choose

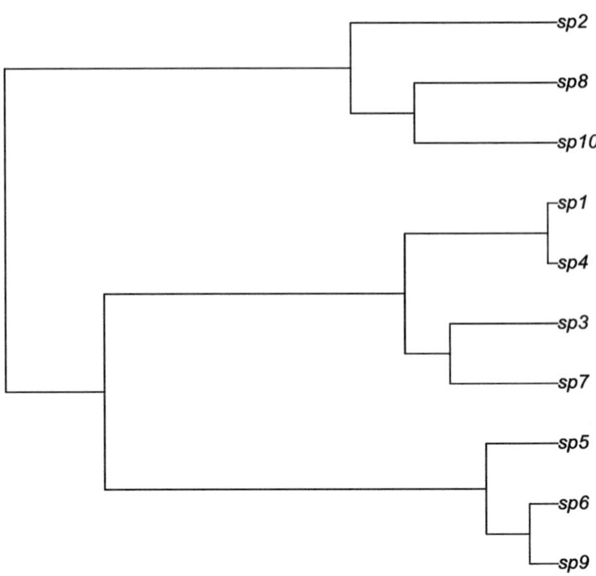

Fig. 6.5 A plot of our example phylogeny with the species names randomized again. Compare with Figs. 6.3 and 6.4

the second option here as it is generally much faster. We start by writing a small function that will take an observed phylogeny, shuffle the names of the species on the phylogeny, and use the randomized phylogeny to calculate the MPD.

```
> my.sample <- read.table("null.example.sample.txt",
sep = "\t", row.names = 1, header = T)

> rand.mpd.fun <- function(x){

    tmp.phylo <- tipShuffle(x)

    ## we use mpd() here, but this function itself
    ## is slow due to its use of for() loops through
    ## rows in the community data matrix. See Chapter
    ## 3 for an alternative and faster approach.
    mpd(my.sample, cophenetic(tmp.phylo))

}
```

Next we can replicate this function ten times to produce ten random MPD values (columns of the output) for each community (rows of the output).

```
> null.output <- replicate(10, rand.mpd.fun(my.phylo))

> null.output
```

When coding null models, it is generally an idea to first utilize a small number of iterations and check the output of the null model. If there is an error in the code or it is slow, you do not want to spend time waiting for thousands of iterations to run to find out. When looking at the null model output matrix of any null that you have

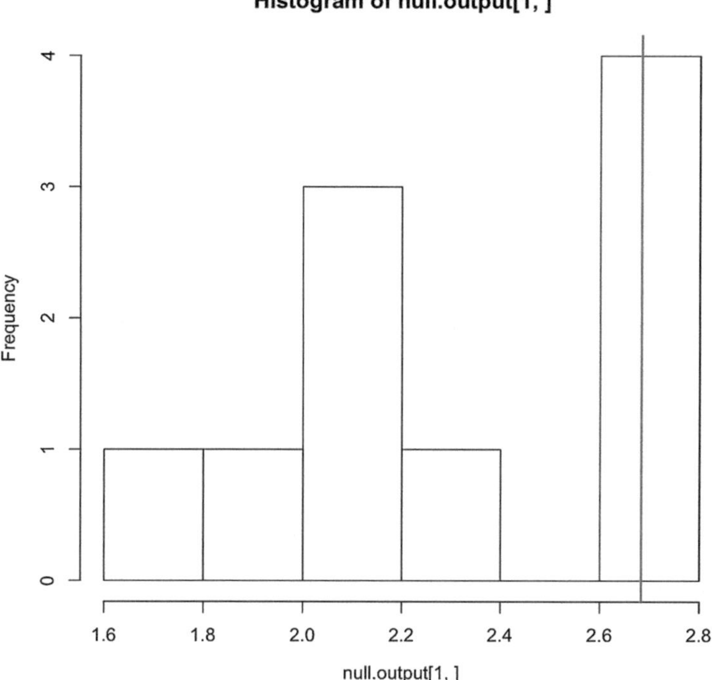

Fig. 6.6 A histogram of our null values calculated from the randomized phylogenetic data. Note that we only ran ten iterations of the null model such that your null distribution may look very different from that displayed and likely is not normally distributed

coded, one should immediately check if the results are indeed random. This could be done by checking the observed values to the observed values in the null model output matrix and by making sure that null values are changing across iterations (values within rows across columns in this instance). If the values in a row are consistent across columns, this indicates that either the code you have written has an error. Another possibility is that the data cannot be randomized to multiple states. For example, if your community contains all species in the meta-community and your phylogeny only contains the species in the meta-community, then you will always have the same MPD value for the community even after shuffling the names of taxa on the phylogeny. Thus it is critical to check the output matrix from a null model particularly one that you have coded de novo.

Now that you have checked the null model output matrix for any obvious errors, plot a histogram to visualize the null distribution. In this example we will look at the null distribution for the first community and therefore will plot the first row of the null model output matrix (Fig. 6.6).

```
> hist(null.output[1, ])
```

Drawing a histogram of the null distribution provides another way of examining whether there appears to be anything odd about the null model. For example, a very low number of values (e.g., all are either an MPD of 125 or 242) or no variance would be warnings that the model is poorly coded or the number of possible random datasets is low, thereby removing your power to reject the null expectation. Recall that you need at least 20 possible states to have enough power to reject a hypothesis using a one-tailed test.

Plotting the null distribution can also be a useful way to visualize how much your observed result does or does not deviate from the null distribution. To do this, simply place a vertical line where your observed value lands in the histogram.

```
> abline(v = mpd(my.sample, cophenetic(my.phylo))[1],
col = "red", lwd = 2)
```

It is possible that your observed value lands far outside the null distribution and you should replot the histogram with an adjusted x-axis that could incorporate your observed value.

Although plotting the null distribution and observed value is useful, we also wish to calculate summary statistics that can be reported and used for quantitatively based inferences. Specifically, we are interested in computing the standardize effect size (S.E.S.) and quantile score. The quantile score, or where the observed value ranks in the null distribution, can be used to calculate a P-value. The S.E.S. is calculated as the observed minus the mean of the null distribution and this value divided by the standard deviation of the null distribution. Here we calculate the S.E.S. for just the first community.

```
> ses.1 <- (mpd(my.sample, cophenetic(my.phylo))[1] -
mean(null.output[1,])) / sd(null.output[1,])

> ses.1
```

Positive S.E.S. values indicate an observed diversity value, MPD in this case, than average null value and negative values indicate an observed diversity value that is lower than average. Recall, that the denominator in these equations standardizes the effect size (numerator) making S.E.S. values more comparable across communities. That said, comparing S.E.S. values across communities or studies may be hindered by differences in species richness causing null distributions to take different shapes. For example, one study may have null distributions that are highly skewed, while another study may have null distributions that are approximately normal. This difference can make comparisons of S.E.S. values difficult. For this reason and others, it is preferable to also compute where the observed value ranks in the null distribution. Calculating such quantile scores can be computed by combining the observed and null values for a community and calculating the rank. In this example, I place the observed value first followed by the null values and then ask for the rank of the first value (i.e., the observed value) in the overall distribution.

```
> rank.1 <- rank(c(observed.mpd, null.output[1,]))[1]
```

This value tells us where the observed value ranks in the overall distribution (i.e., the quantile score). This value can be utilized to calculate a *P*-value by dividing it by the number of null model iterations plus one.

```
> p.val.1 <- rank.1 / 11
```

The value reported is the *P*-value for where the observed lands in the null probability distribution. In this example we have only 10 iterations of the null model and one observed value making 11 total values and therefore a denominator of 11. Remember that if your study uses 999 randomizations, the number 11 would be changed to 1,000. If the test were a one-sided test of whether the MPD value observed was significantly lower than expected, we would require a *P*-value less than or equal to 0.05. If the test were a one-sided test of whether the MPD value observed was significantly higher than expected, we would require a *P*-value greater than or equal to 0.95. If the test were two sided, we would require *P*-values less than or equal to 0.025 or greater than or equal to 0.975.

Instead of calculating the S.E.S. or *P*-value for each community one by one, the `apply()` function can be used to calculate these values for all communities simultaneously.

```
> ses.all <- (mpd(my.sample, cophenetic(my.phylo)) -
apply(null.output, MARGIN = 1, mean)) /
apply(null.output, MARGIN = 1, sd)
```

Here we use the `apply()` function to calculate the mean and standard deviation of the values in each row of the null model output matrix. The same approach can be utilized for calculating the *P*-values.

```
> p.val.all <- apply(cbind(mpd(my.sample,
cophenetic(my.phylo)), null.output), MARGIN = 1,
rank)[1,] / 11
```

Now that we have seen to code a null model randomizing the names of taxa on a phylogeny for the MPD metric, let us quickly consider how to code this same randomization for the PD metric. The PD metric is being used as a second example simply because the output from the `pd()` function is slightly different.

As with the previous example, we want to start with determining what the output from the function of interest look like so we know how to generate a null model output matrix.

```
> observed.pd <- pd(my.sample, my.phylo, include.root
= FALSE)
> observed.pd
```

Unlike the vector output from the `mpd()` function, the output from the `pd()` function is matrix with two columns. The first column reports the PD value and the second column reports the species richness (SR) value and each row represents a community in the meta-community. Since the species richness is not being manipulated in our randomizations and because it is not of immediate interest for this randomization study we do not wish to have this information in our output matrices. We therefore can generate a randomization function reporting out just the PD values.

```
> rand.pd.fun <- function(x){

    tmp.phylo <- tipShuffle(x)

    pd(my.sample, tmp.phylo)[,1]

}
```

Next we can replicate this function ten times to produce ten random MPD values (columns of the output) for each community (rows of the output).

```
> null.output <- replicate(10, rand.pd.fun(my.phylo))

> null.output
```

We can now use the `apply()` function as before to calculate our S.E.S. values and *P*-values.

```
> ses.all <- (observed.pd - apply(null.output, MARGIN
= 1, mean))/apply(null.output, MARGIN = 1, sd)
```

Here we use the `apply()` function to calculate the mean and standard deviation of the values in each row of the null model output matrix. The same approach can be utilized for calculating the *P*-values.

```
> p.val.all <- apply(cbind(observed.pd, null.output),
MARGIN = 1, rank)[1,] / 11
```

The above has demonstrated how to write a simple randomization of names of taxa on a phylogenetic tree and how this can be used to calculate the S.E.S. and *P*-value for phylogenetic diversity metrics such as MPD and PD. In Sect. 6.7 we will examine functions that are already coded in R that can be utilized for the calculation of these values for most phylogenetic diversity metrics.

6.5.2 *Constrained Randomizations*

The great advantage of randomizing the names of taxa on the phylogeny is that all observed patterns in the community data matrix are fixed. Thus, the total abundance of species with and across communities, the occupancy rates of species across communities, the species alpha and beta diversity, and patterns of the spatial contagion of species (i.e., dispersal limitation) are all fixed. These benefits are generally enough for me to strictly adhere to using the randomizing of names of taxa on the phylogeny for all analyses of phylogenetic alpha and beta diversity. Though, some cases it may be necessary to further constrain this name shuffling null model. For example, Hardy [126] highlighted that type I error rates could be inflated using a name shuffling null model if there is phylogenetic signal in abundance or occupancy rates, that is, when closely related species tend to be equally abundant or have similar occupancy rates. In Chap. 7 we will discuss how to calculate the phylogenetic signal in trait, abundance and occupancy data. In cases where there is phylogenetic signal in abundance or occupancy rate, Hardy [126] suggests placing species into bins based on their global abundances (i.e., their total abundance across all communities) or the number of communities they occur in (i.e., their occupancy rates in the meta-community). The problem with any analysis that relies on the binning of data is that it requires the investigator or arbitrarily set the limits to the individual bins or to determine bin limits based on some aspect of biology envisioned from the perspective of a human. The results of investigations that bin data are therefore often very sensitive to the decisions made by the investigator and are difficult to compare across separate studies. Unfortunately, the bins chosen for the null proposed by Hardy [126] are indeed arbitrary with no set rule to follow across studies, but I will still provide an example.

Binning and shuffling names based on their abundances fixes the distribution of occupancy rates across species but does not maintain the occupancy rates of individual species. Indeed, the change in occupancy rate could change dramatically. For example, if species A has 48 individuals spread across 4 communities and species B has 47 individuals all located in 1 community, a binning and shuffling based on abundance would place these two species in the same bin and randomize them as if they are equal. Similarly, binning and shuffling by occupancy rate could result in the shuffling of species with very different global abundances. Thus, as we have seen before, even in "constrained" null models designed to reduce type I error rates, observed properties of the data may be inadvertently randomized when constraining other observed aspects of the data. Great care should therefore be taken to understand exactly what is and is not being directly or indirectly randomized in all null models.

The constrained shuffling null model of Hardy [126] can be implemented in R using the following code. I will bin the data by abundance arbitrarily since there is no set rule for bin size. The bin size (K) is used to set the limits of the abundance bins using a geometric series. A geometric series can be quickly generated using the `gseq()` function in the *Rsundials* R package.

```
> library(Rsundials)
```

The `gseq()` function produces a geometric series using a starting value, an ending value, and an exponent value. The starting value here is the minimum global abundance from our community data matrix, the ending value is the maximum global abundance, and the exponent is the Hardy K value which we can define arbitrarily as 3.

```
> hardy.K <- 3

> abund.bins <- gseq(from = min(colSums(my.sample)),
to = max(colSums(my.sample)), by = hardy.K)

> abund.bins
```

The values in the output are the lower bound of each bin. Thus, the present output has four values indicating a total of four abundance bins. The next step is to bin each species in the community data matrix based on where its global abundance falls in the bins we just defined. The global of all species can be calculated by using `col-Sums()`. The global abundances and the bin definitions can be used to assign a bin to each species using the `findInterval()` function.

```
> assigned.bins <- findInterval(colSums(my.sample) ,
abund.bins)

> assigned.bins
```

The vector output is in the same order as the species names in the community data matrix and we can assign them as follows.

```
> names(assigned.bins) <- names(colSums(my.sample))

> assigned.bins
```

Now that we have names assigned to the vector we can bin species, the names of the vector, based on the numerical values in the vector.

```
> split.bins <- split(names(assigned.bins),
assigned.bins)

> split.bins
```

The result is a list with one element per bin containing all of the species names with that have global abundances falling inside that bin. The next goal is to shuffle names within each of these elements. We will first write a simple function we will call `hardy.shuffle()` that takes an input vector of values and shuffles them.

```
> hardy.shuffle <- function(x){

    sample(x, length(x), replace = F)

}
```

We can now apply this function to each individual element in the list to shuffle the names in each bin using `lapply()`.

```
> lapply(split.bins, hardy.shuffle)
```

The names of the species within each bin have now been shuffled. The goal now is to combine the names. It is critical that we do not simply combine the names in the order they are displayed. Rather, the shuffled names must be pieced together in the same order as the bins in the `assigned.bins` object. In other words, using the shuffled names in the first bin we want to assign the first name to the first occurrence of the first bin in the `assigned.bins` object, the second name to the second occurrence of the first bin, and so on. This can be accomplished using the `unsplit()` function using our split and randomized object and the information from the original values in the `assigned.bins` object.

```
> unsplit(lapply(split.bins, hardy.shuffle),
  assigned.bins)
```

We now see that we have recombined the names into a single vector, but they are in a randomized order within individual bins. To see this we can inspect the order of the names and the bin numbers in the original object.

```
> assigned.bins
```

Now that we know how to perform a single iteration of the null model, we can scale it up. We will construct a simple example null model with nine iterations. We first use the `replicate()` function to shuffle and recombine our binned names.

```
> null.names <- replicate(9,
  unsplit(lapply(split.bins,hardy.shuffle),
  assigned.bins))
```

We now have a matrix with the same number of rows as we have species in the community data matrix and one column per iteration of the null model.

```
> null.names
```

We will now sequentially assign these names to the community data matrix, but we do not wish to alter the original data matrix so we can make a duplicate community data matrix that we can alter.

```
> tmp.sample <- my.sample
```

A simple function can now be written that assigns new column names to our temporary community data matrix and calculates the PD using Faith's Index and our original phylogeny and the data matrix containing the given species names.

```
> hardy.pd.null <- function(x){

    colnames(tmp.sample) <- x

    pd(tmp.sample, my.phylo)[,1]

}
```

The above calculation could easily be changed to incorporate other phylogenetic diversity metrics. For example, the mean pairwise phylogenetic distance (MPD) could be calculated as follows:

```
> hardy.mpd.null <- function(x){

    colnames(tmp.sample) <- x

    mpd(tmp.sample, cophenetic(my.phylo))

}
```

The replicated name shuffling and function above can be now be used in an `apply()` to rapidly calculate the random PD value for each community and set of shuffled names where we take the shuffled names in each column and apply the function that assigns those names to the community data matrix and calculates PD.

```
> apply(null.names, MARGIN = 2,  hardy.pd.null)
```

The result is a matrix with one column per iteration of the null model, 9 in this scenario, and one row per community in the community data matrix. To a full null model, one could easily change the 9 in the above `replicate()` function to 999, 9999, etc.

6.6   Randomizing Functional Trait Data

The randomization of functional trait data for analyses of alpha and beta FD generally involve the shuffling of species names on the trait data matrix. While this may be considered an unconstrained null model, it does constrain the co-variances and overall phenotypes while changing the name of the species that has that phenotype. In the next subsection we will demonstrate how to code this simple null model. The constrained trait null model that we will discuss in the subsection after that discusses how to constrain a null model for nearest neighbor distances by the observed range. Specifically, the range of a trait is generally considered to indicate the degree to which the abiotic environment filters species or traits into a habitat where biotic interactions should be stronger and therefore influence the nearest neighbor. By not controlling the observed range in randomizations for nearest neighbor indices, it is possible that results of the null model may be biased toward finding smaller than expected nearest neighbor distances. First, though, let us quickly consider a less constrained name shuffling null model.

6.6.1   *Unconstrained Randomizations*

The least constrained null model that isn't ridiculous (i.e., shuffling all trait values in each column independently) is to simply shuffle the row names on the trait matrix. This maintains the observed patterns of trait co-variance and overall phenotypes. This null is therefore somewhat analogous to the name shuffling null used for phylogenies. To shuffle the names on a trait matrix we can first read in the trait matrix and then make a duplicate of that matrix that we can randomize.

```
> traits <- read.table("null.example.traits.txt", sep
= "\t", header   T, row.names = 1)

> rand.traits <- traits

> rand.traits
```

Next we simply randomly sample all row names in the trait matrix with replacement effectively shuffling their order.

```
> replicate(5, sample(rownames(rand.traits),
length(rownames(rand.traits)), replace = F))
```

To utilize this approach to write a null model for your favorite FD metric, we first make a function for the randomization. In this case we will use a mean pairwise trait distance.

```
> trait.shuffle.funk <- function(x){

    x <- x[colnames(my.sample), ]

    rownames(x) <- sample(rownames(x),
    length(rownames(x)), replace = F)

    mpd(my.sample,
    as.matrix(dist(x[colnames(my.sample), ])))

}
```

This function can now be replicated multiple times (i.e., output columns) to produce a null distribution for each community (i.e., output rows).

```
> replicate(10, trait.shuffle.funk(traits))
```

To calculate the S.E.S. and *P*-value for each community, you could use the output from this `replicate()` function in the same way we used the `replicate()` output in Sect. 6.5.1.

6.6.2   Constrained Randomizations

The discussion of constrained randomizations of trait data has generally been limited to a single approach. Specifically, the tradition of simultaneously analyzing the trait range or volume and the mean nearest trait neighbor distances in assemblages has been criticized. The basis of the criticism is that functional ecologists typically conceptualize the assembly of species in a community as first a process of abiotic filtering at larger spatial scales that should limit the range or volume of functions in a community. Within this limited range biotic interactions or finer scale abiotic gradients dictate the functional similarity of species. If null models shuffling species names on the trait matrix are performed on both the range or volume and the nearest neighbor distance metrics it may be expected that we under-estimate the observed nearest neighbor distance in relation to the null expectation. Specifically, if the null model utilizes all species in the region, even those that fall outside the range or volume of traits observed in the assemblage, the nearest neighbor distances will seem much lower than expected for those communities that have small ranges or volumes. Though, if the expectation for the nearest neighbor distances was based upon a null model that only selected species from those that had traits within the observed range or volume of traits this bias might be removed [127, 128]. Thus, for those conceptualizing trait-based community assembly as first a process where abiotic filtering constrains the range or volume of traits in a community and second a process where local scale abiotic and biotic interactions determine the spacing of species in the given range or volume of trait space, a more constrained null modeling approach may be preferred.

To accomplish the constrained null modeling process outlined in the above paragraph there is no change in how one would analyze the deviation of the observed range or volume of traits in an assemblage versus that expected using a null model that shuffles the names of species on the trait data matrix. The mean nearest neighbor null model however must be modified. Specifically, the names of species can no longer be shuffled for all species in the trait data matrix. Rather, we must first trim the observed trait matrix to only contain those species that have trait values that fall within the uni-variate range of the trait being analyzed or the multi-variate volume of the traits being analyzed. It is recommended that even in those cases where you are analyzing a single trait, if you have information for multiple traits you should constrain the trait data matrix (i.e., the species pool) by the multi-variate trait volume observed to further reduce Type I error. Once the trait data matrix has been constrained to only include species that fall within the observed trait volume, we can shuffle the names on this constrained trait matrix and run the null model. In the following, I provide a function that utilizes the three traits in our trait data matrix to calculate the mean nearest functional neighbor for a single community (x). This function can then be applied to all communities and replicated 999 times to generate a null distribution. I provide comments throughout the function so that it is clear what is being done and so you can determine how to alter the function to suit your particular needs.

```r
> shuff.constrain.nn <- function(x){

    ## Calculate the maximum and minimum value for
    ## trait 1 in community
    max.sam.1 <- max(traits[names(x[x > 0]), 1])

    min.sam.1 <- min(traits[names(x[x > 0]), 1])

    ## Calculate the maximum and minimum value for
    ## trait 2 in community
    max.sam.2 <- max(traits[names(x[x > 0]), 2])

    min.sam.2 <- min(traits[names(x[x > 0]), 2])

    ## Calculate the maximum and minimum value for
    ## trait 3 in community
    max.sam.3 <- max(traits[names(x[x > 0]), 3])

    min.sam.3 <- min(traits[names(x[x > 0]), 3])

    ## Get the names of all the species in your trait
    ## data that are less or equal to the maximum
    ## trait 1 value observed in community and names
    ## of species greater than or equal to the
    ## observed minimum trait 1 value in community.

    tr.1.a <- names(traits[traits[ ,1] <= max.sam.1,
    1])

    tr.1.b <- names(traits[traits[ ,1] >= min.sam.1,
    1])

    ## Repeat the above for trait 2.
    tr.2.a <- names(traits[traits[ ,2] <= max.sam.2,
    2])

    tr.2.b <- names(traits[traits[ ,2] >= min.sam.2,
    2])

    ## Repeat the above for trait 3.
    tr.3.a <- names(traits[traits[ ,3] <= max.sam.3,
    3])

    tr.3.b <- names(traits[traits[ ,3] >= min.sam.3,
    3])

    ## Intersect all of the names from above such
    ## that the only remaining names are species that
    ## fall within the three dimensional range for
    ## the community. This is the new species pool.
    pruned.names <- Reduce(intersect, list(tr.1.a,
    tr.1.b, tr.2.a, tr.2.b, tr.3.a, tr.3.b))

    ## Prune sample to only include species in new
    ## species pool.
    pruned.sam <- my.sample[ , pruned.names]

    ## Prune trait matrix to only include species in
    ## new species pool.
    pruned.matrix <- traits[pruned.names, ]

    ## Shuffle the species names on the pruned trait
    ## matrix
    rownames(pruned.matrix) <-
    sample(rownames(pruned.matrix),
    length(rownames(pruned.matrix)), replace = F)

    ## Calculate mean nearest neighbor distance.
    mntd(pruned.sam, as.matrix(dist(pruned.matrix)),
    abundance.weighted = F)

}
```

We now replicate the above 999 times getting a random value for each community each time using a constrained range. When replicating the `shuff.constrain. nn()` function once, we see that only the diagonal values are of interest. This is because the `apply()` function goes through each row but we report values for each community. Thus we only want the values for the community we are constraining the pool by. This means that we would like the diagonal of every one of the 999 layers in the output array.

```
> apply(replicate(999,apply(my.sample, 1,
shuff.constrain.nn)), 3, function(x){ diag(x) })
```

We have now successfully generated a null model for mean nearest functional neighbor analyses that is constrained by the observed volume defined by the observed minimum and maximum of multiple trait values in a community. The above code could be modified to include one to many traits and other metrics such as the standard deviation of the nearest neighbor distances in the assemblage. For example, if you only had two traits in your trait data matrix, you would remove all lines of code concerning trait 3, or if you had four traits, you would copy and paste the lines of code concerning trait 3 and change the 3's to 4's. Further, if you wanted to constrain the null model to the observed multivariate trait volume, but only wanted the mean nearest neighbor distance for a single trait, you could change the last line in the `shuff.constrain.nn()` function from using `as. matrix(dist(pruned.matrix))` to using `as.matrix(dist(pruned. matrix[,1]))` for trait 1.

The rationale for the above-constrained null modeling approach is that the range of trait values is likely first filtered by the abiotic environment, and within this range, species are again filtered based upon their similarity. By not constraining the null model by the observed range, the nearest neighbor results may be biased toward finding lower than expected mean nearest neighbor distances. As with any other constrained null model, however, this approach takes a little more computational time and may reduce the number of random possibilities for a given dataset to a point at which there is no statistical power to reject the null hypothesis. As such, when using this null modeling approach, the researcher must be clear that they are making the biological assumption that the range or volume of traits is first filtered during assembly prior to the interactions that underlie the interactions that determine the similarity of species in that constrained space. Further, the researcher must be cognizant that in many cases the observed range may be so small as to restrict them from having any power to reject the null hypothesis regarding mean nearest neighbor distances.

6.7 Null Models for Phylogenetic and Functional Alpha Diversity

In this section we will quickly run through how to perform null models for the major classes of phylogenetic and functional diversity metrics and to calculate the summary statistics. In each example, I will first demonstrate how to generate our own code for the null model and then how to utilize existing functions in the R package *picante*. In most cases, the two approaches will result in similar computation times, but if you are working with a large dataset, it may be noticeably faster to utilize the code we write below and not that in the *picante* package.

We start by performing an independent swap null model for Faith's Index. First, we will write a simple function that first randomizes a community data matrix once and then calculates a random Faith's Index value for each community.

```
> pd.is <- function(x){

    pd(randomizeMatrix(x, null.model =
    "independentswap"), my.phylo )[,1]


}
```

We can now use this function to generate 999 random Faith's Index values for each community where each community is a row in the output and each random value is in its own column for each row. We will bind to the left of this matrix the observed Faith's Index for each community. The result is a matrix with one row per community and 1,000 columns the first of which is the observed value and the following 999 are the random values.

```
> obs.null.output <- cbind(pd(my.sample,
my.phylo)[,1], replicate(999, pd.is(my.sample)))
```

From this large matrix we can easily quantify the rank of the observed value in the null distribution for each community using an `apply()` function. Recall that the rank can be used to estimate the P-value. In this instances we may wish to perform a two-tailed test where we want to test whether the Faith's Index value is larger or smaller than expected. If it is larger than expected, we would need to observe a rank greater than or equal to 975 (i.e., P-value$=1-975/1,000$). If it is smaller than expected, we would need to observe a rank less than or equal to 25 (i.e., P-value$=25/1,000$).

```
> obs.rank <- apply(obs.null.output, 1, rank)[,1]

> obs.rank
```

We can also use the matrix of observed and null values to quantify the standardized effect size (S.E.S.) for each community. Recall, that to calculate a S.E.S., we subtract the mean of the null distribution from the observed value and then divide this by the standard deviation of the null distribution. Therefore positive values are indicating the observed value is higher than the average random expectation and negative values are indicating the observed value is lower than the average random expectation.

```
> ses.value <- (obs.null.output[,1] -
apply(obs.null.output, 1, mean)) / -
apply(obs.null.output, 1, sd)

> ses.value
```

If you would rather not use the above code, you could utilize the `ses.pd()` function in the *picante* package using the "independentswap" null model.

```
> ses.pd(my.sample, my.phylo, null.model =
"independentswap", runs = 999, iterations = 1000)
```

The first column in the output is the species richness, the second column is the observed Faith's Index, the third column is the mean of the null distribution, the fourth column is the standard deviation of the null distribution, the fifth column is the rank of the observed in the null distribution, the sixth column is the S.E.S. (called "z" here), the seventh column is the estimated *P*-value and the final column is the number of randomizations performed.

We can run the same analysis with the exception of not randomizing the community data matrix and randomizing the names of species on the phylogeny. We can perform this ourselves writing the following simple function and replicating it 999 times.

```
> pd.shuff <- function(x){

    pd(my.sample, tipShuffle(x)[,1]

}


> obs.null.output <- cbind(pd(my.sample,
my.phylo)[,1], replicate(999, pd.shuffle(my.phylo)))
```

The output matrix could be used to calculate the S.E.S. values and *P*-values as demonstrated in the preceding example. The analysis could also be performed using the `ses.pd()` function and the "taxa.labels" null model in *picante*.

```
> ses.pd(my.sample, my.phylo, null.model =
"taxa.labels", runs = 999, iterations = 1000)
```

The output columns have the same meaning as described above with the exception that a different null modeling approach has been utilized.

Now that we have this basic structure down, below I provide simple code for how you could code your own null models using the independent swap or name shuffling null models in the context of a `replicate()` function and I also provide the *picante* function that utilizes a `for()` loop approach.

The mean pairwise distance (MPD) for phylogenetics can be calculated as follows though the `cophenetic(my.phylo)` object could be replaced with a trait distance matrix for functional calculations.

```
> mpd.is <- function(x){

    mpd(randomizeMatrix(x, null.model =
    "independentswap"), cophenetic(my.phylo))

}


> obs.null.output <- cbind(mpd(my.sample,
cophenetic(my.phylo)), replicate(999,
mpd.is(my.sample)))
```

This can be calculated in *picante* using `ses.mpd()` and the "independentswap" null.

```
> ses.mpd(my.sample, cophenetic(my.phylo), null.model =
"independentswap", abundance.weighted = F,  runs =
999, iterations = 1000)
```

A null model for MPD that utilizes a shuffling of species names on the phylogeny can be coded.

```
> mpd.shuff <- function(x){

    mpd(my.sample, cophenetic(tipShuffle(x)))

}
```

If you wished to change the above function to focus on traits instead of a phylogeny you could do the following where the x object is a trait data matrix with row names being species names.

```
> mpd.shuff.trait <- function(x){

    rownames(x) <- sample(rownames(x))
    mpd(my.sample, as.matrix(dist(x)))

}

> obs.null.output <- cbind(mpd(my.sample,
cophenetic(my.phylo)), replicate(999,
mpd.shuffle(my.phylo)))
```

This can be calculated in *picante* using `ses.mpd()` and the "taxa.labels" null.

```
> ses.mpd(my.sample, cophenetic(my.phylo), null.model =
"taxa.labels", abundance.weighted = F,  runs = 999,
iterations = 1000)
```

The mean nearest taxon distance (MNTD) (a.k.a. the mean nearest neighbor distance) can be calculated using an independent swap null model as follows, and as in the above MPD-independent swap null model, a trait data distance matrix could be used to replace the phylogeny to calculate the mean nearest functional neighbor distance.

```
> mntd.is <- function(x){

    mntd(randomizeMatrix(x, null.model =
    "independentswap"), cophenetic(my.phylo))

}

> obs.null.output <- cbind(mntd(my.sample,
cophenetic(my.phylo)), replicate(999,
mntd.is(my.sample)))
```

This can be calculated in *picante* using `ses.mntd()` and the "independentswap" null.

```
> ses.mntd(my.sample, cophenetic(my.phylo), null.model
= "independentswap", abundance.weighted = F,  runs =
999, iterations = 1000)
```

A null model analysis for MNTD can be calculated using a shuffling of species names on the phylogeny using the following function and replicating it.

```
> mntd.shuff <- function(x){

    mntd(my.sample, cophenetic(tipShuffle(x)))

}
```

If you wished to change the above function to focus on traits instead of a phylogeny you could do the following, where the x object is a trait data matrix with row names being species names.

```
> mntd.shuff.trait <- function(x){

    rownames(x)  = sample(rownames(x))
    mntd(my.sample, as.matrix(dist(x)))

}

> obs.null.output <- cbind(mntd(my.sample,
cophenetic(my.phylo)), replicate(999,
mntd.shuffle(my.phylo)))
```

This can be calculated in *picante* using `ses.mntd()` and the "taxa.labels" null.

```
> ses.mntd(my.sample, cophenetic(my.phylo), null.model
= "taxa.labels", abundance.weighted = F,  runs = 999,
iterations = 1000)
```

The above text performs null model analyses for metrics that were coded phylogenetic diversity analyses in *picante* that can also be used for functional diversity analyses. However, we also know that additional metrics of functional diversity are available in the *FD* package, but these metrics do not have null model analyses available in this package. Thus we must write our own null models in R. In the following I provide simple null model analyses using an independent swap null and name shuffling null for the FD is metric, but the $FDis could be replaced in this code to $FDiv, $FEve, or $FRic for those metrics.

First we generate a function that randomizes our community data matrix using an independent swap null model. Then we calculate the FDis metric using the `dbFD()` function in the *FD* package.

```
> dbfd.is <- function(x){

    dbFD(traits, randomizeMatrix(my.sample,
    null.model = "independentswap"))$FDis

}
```

Next we replicate the above function 999 times and combine the observed output with the 999 random values per community.

```
> obs.null.output <- cbind(dbFD(traits,
my.sample)$FDis, replicate(999, dbfd.is(traits)))
```

To generate null FDis values for each community by shuffling the names of species on the trait data matrix, we can first write this simple function to shuffle names and then calculate a null FDis value using the dbFD() function.

```
> dbfd.shuff <- function(x){

    rownames(x) <- sample(rownames(x),
    length(rownames(x)), replace = F)

    dbFD(x, my.sample)$FDis

}
```

We now replicate this function 999 times and combine it with the observed output.

```
> obs.null.output <- cbind(dbFD(traits,
my.sample)$FDis, replicate(999, dbfd.shuff(traits)))
```

At this point we have now coded null models for phylogenetic and functional alpha diversity and covered those functions in R that are already coded for similar analyses. The above code is flexible enough to be re-tooled for alternative metrics of phylogenetic or functional alpha diversity that you may come across in R or that you have generated yourself. As always it is good practice to first test your code on a small dataset with a few iterations of the null model prior to applying it to your entire dataset.

6.8   Null Models for Phylogenetic and Functional Beta Diversity

In the previous section we discussed how to implement null model analyses for phylogenetic and functional alpha diversity. In many instances we saw that there were pre-existing R functions to implement these analyses. The case is not the same for phylogenetic and functional beta diversity where code to generate null distributions to quantify a standardized effect size (S.E.S.) and P-value do not exist. In some instances, there is an R function to generate a null distribution (see phylosor.rnd() function in the *picante* package), but these functions are generally very slow and do not calculate the S.E.S. or P-value for you. Thus, in this section we will generate our own null model code that can be used to calculate the S.E.S. and P-value for pairwise and nearest neighbor metrics of phylogenetic and functional beta diversity. We will use the same general approach as we used for alpha diversity with one main exception. Alpha diversity analyses result in one value per community (i.e., vector or one column matrix with one value per community), but beta diversity analyses provide a dissimilarity between each pair of communities (i.e., a matrix where the number or rows and columns is equal to the number of communities). The pairwise nature of the output requires that we utilize arrays to store the null results with each "layer" in the z-dimension of the three-dimensional array contains the beta diversity between all pairs of communities for a single iteration of the null model. This complicates the calculation of S.E.S. values and P-values slightly and takes longer to compute, but the general approach is the same.

We begin by writing a function that randomizes the community data matrix once using an independent swap approach and calculates the pairwise phylogenetic dissimilarity (D_{pw}) value between each pair of communities.

```
> comdist.is <- function(x){

    as.matrix(comdist(randomizeMatrix(x, null.model =
    "independentswap"), cophenetic(my.phylo),
    abundance.weighted = F))
}
```

The above function could be modified to utilize a trait distance matrix instead of the phylogenetic tree to calculate the functional D_{pw}.

```
> nulls <- replicate(999, comdist.is(my.sample))
```

We can now replicate our function 999 times to generate our null distribution for each community comparison. Note that this calculation can be time intensive for large datasets, but for the present example dataset, the computational time will be short. The output `nulls` object will be a three-dimensional array, where the row and column numbers are equal to the number of communities and the number of levels in the z-dimension is 999. Thus, the null distribution of the D_{pw} values between two communities can be described using the third dimension. For example, to plot the null distribution of D_{pw} values between community 1 and community 2, we can use the following code.

```
> hist(nulls[1, 2, ])
```

In order to calculate the S.E.S. value for each pair of communities, we need to compute a mean and standard deviation of the null distribution in the z-dimension. We can do this simply using two `apply()` functions.

```
> nulls.means <- apply(nulls, c(1:2), mean, na.rm = T)

> nulls.sds <- apply(nulls, c(1:2), sd, na.rm = T)
```

Next we can calculate the observed D_{pw} values for our communities using the original data.

```
> obs <- as.matrix(comdist(my.sample,
cophenetic(my.phylo), abundance.weighted = F))
```

We now have the observed D_{pw} value for each pair of communities and the mean and standard deviation of the null D_{pw} values for each pair of communities. We can use this information to quickly calculate the S.E.S. values.

```
> ses <- (obs - nulls.means) / null.sds
```

Next we need to combine the observed data with the array of null values such that the first level in the new array contains the observed values and the following 999 levels contain the random values. We can do this by binding the observed matrix to the random array using the *abind* package.

```
> install.packages("abind")

> library(abind)

> obs.nulls <- abind(obs, nulls)
```

Now we calculate the rank of the values in each layer. We will do this by outputting an array with the value in each cell is the rank of the D_{pw} value in that cell in the null distribution in the z-dimension. We can do this using two `apply()` functions one for ranking the values and one to transpose the results such that they align with the original array.

```
> temp.rank <- array(dim = dim(obs.nulls),
t(apply(apply(obs.nulls, c(1, 2), rank), 3, t)))
```

Although we might be interested in where some null values are ranked in the distribution, for calculating *P*-values we are only concerned where the observed values are ranked. Thus we will only want the values in the top, or first, layer in the ranked output.

```
> temp.rank[ , ,1]
```

As with all of the other null model analyses where we have used 999 randomizations and a two-tailed test, rank values less than or equal to 25 indicate an observed value that is lower than that expected and values of 975 or greater indicate an observed value that is higher than expected.

The above can be repeated using a null model that shuffles names on the phylogeny or trait matrix instead of randomizing the community data matrix with an independent swap null model. I will demonstrate this here using a shuffling of names on a phylogeny.

```
> comdist.shuff <- function(x){

        as.matrix(comdist(my.sample,
        cophenetic(tipShuffle(x)), abundance.weighted =
        F))

}
```

We now replicate this randomization 999 times to generate an array of null D_{pw} values between each pair of communities.

```
> nulls <- replicate(999, comdist.shuff(my.phylo))
```

The output `nulls` object could be used to calculate the S.E.S. values and P-values using the same approach used above for the community data matrix D_{pw} null.

Once we have learned how to calculate the independent swap and name shuffling null models for D_{pw}, they can be easily modified and applied to generate the same type of null models for the nearest neighbor dissimilarity (D_{nn}) between communities. Specifically, the independent swap approach for phylogenetic D_{nn} can be implemented as:

```
> comdistnt.is <- function(x){

        as.matrix(comdistnt(randomizeMatrix(x, null.model
        = "independentswap"), cophenetic(my.phylo),
        abundance.weighted = F, exclude.conspecifics =
        F))

}

> nulls <- replicate(999, comdistnt.is(my.sample))
```

The name shuffling null for phylogenetic D_{nn} can be implemented as:

```
> comdistnt.shuff <- function(x){

        as.matrix(comdistnt(my.sample, tipShuffle(x),
        abundance.weighted = F, exclude.conspecifics =
        F))

}

> nulls <- replicate(999, comdistnt.shuff(my.sample))
```

While the above code focuses on the presence–absence weighted D_{pw} and D_{nn} metrics, the abundance-weighted versions, D_{pw}' and D_{nn}' respectively, can be calculated by simply changing the abundance.weighted option to "T." Further, the above approaches can be easily modified to calculate the functional trait metrics using a trait distance matrix in each. Specifically, the functions would be modified as:

```
> comdistnt.is.traits <- function(x){

    as.matrix(comdistnt(randomizeMatrix(x, null.model
    = "independentswap"), as.matrix(dist(traits)),
    abundance.weighted = F, exclude.conspecifics =
    F))

}

> comdistnt.shuff.traits <- function(x){

    rownames(x) <- sample(rownames(x))

    as.matrix(comdistnt(my.sample, x,
    abundance.weighted = F, exclude.conspecifics =
    F))

}
```

As with all null modeling studies you will be asked to explain your choice of one null model over another. I typically prefer to use name shuffling for my analyses of phylogenetic and functional alpha and beta diversity. In the case of alpha diversity, the use of other constrained null models like the independent swap that focus on randomizing the community data matrix and not the phylogeny or trait data may be a reasonable option. However, my personal viewpoint is that independent swap null models should not be used for phylogenetic and functional beta diversity analyses. This is because the independent swap null model does not maintain the spatial structure of species in the system. In other words any spatial contagion or dispersal limitation is not maintained in the null community data matrices. Given that those investigating spatial beta diversity are often interested in the degree to which space or dispersal limitation or the environment determine community turnover, it is problematic that the null community data matrices do not constrain the observed spatial patterns and likely inflate your bias toward finding lower than expected turnover. With respect to the temporal turnover of communities the independent swap also has problems because the number of individuals of a species is not constrained. Thus an individual tree that is present in census 1 and census 2 could have, in theory, two different species names in an independent swap null model and such a scenario is clearly not biologically reasonable. The shuffling of names on the other hand assures that the observed spatial patterns of species in your system are maintained allowing for clearer inferences regarding the role of spatial and environmental gradients and the analyses do not suffer from the biologically unreasonable possibility that individuals surviving from time 1 to time 2 change from one species to another.

6.9   Conclusions

Here we have considered the conceptual underpinnings of null model analyses in phylogenetic and functional diversity analyses and how relatively unconstrained and constrained null models can be implemented in R. Null models can be a powerful tool for such analyses, but it is critical that a researcher understands and can articulate why they chose a particular null model for their analyses. It is particularly important that one recognizes what is and is not fixed in their null model randomizations and how this information does or does not link up with the general goal of null modeling which is to fix all of the observed patterns except the one pattern of interest. Often it will be difficult to achieve this ideal because so many patterns in ecology co-vary tightly, but this is not an excuse for having a highly unconstrained null model or worse no null model at all. Computational power is potentially the only real excuse for a highly unconstrained null model or no null model, but this obstacle is quickly eroding and R in particular makes it particularly easy to code your own null models. In those instances you do code your own null models, attempt to avoid for() loops wherever necessary and always test your code on multiple small datasets to be sure that your code is actually doing what you hoped it would.

6.10   Exercises

1. Write a null model for PhyloSor or UniFrac. Run it 99 times, output the observed result, the mean of the null distributions, the standard deviation of the null distributions, the standardized effect sizes, and the *P*-values.
2. Write a null model for a Jaccard's Index (use the function `vegdist(my.sample, method = "jaccard")`) in the *vegan* package), run it 99 times on the example dataset, and output the standardized effect sizes values using an Independent Swap null model.
3. Repeat number 2 above, but calculate the Checkerboard Score (C-Score) of co-occurrence for all species pairs in your meta-community. The C-Score can be calculated using the function `C.score()` in the *bipartite* package.
4. Using the results from number 3 above, plot the null distribution of C-Scores as a histogram. Then plot your observed value as a red vertical line on the histogram.

Chapter 7
Comparative Methods and Phylogenetic Signal

7.1 Objectives

The objectives of this chapter are to consider trait data in the context of phylogenetic information. We will begin by discussing how to quantify the relationships between the traits of species while accounting for the nonindependence of species. We will then explore how to quantify the degree to which variation in traits reflects phylogenetic nonindependence. Throughout we will discuss multiple approaches, but keep in mind that not all approaches are equal. Some approaches that will be presented are now rarely used due to their documented weaknesses. Nonetheless we will cover these approaches to provide some breadth, background, and context.

7.2 Trait Correlations

Biologists have long recognized that closely related species are generally more similar to one another than they are to more distantly related. This is often termed phylogenetic conservatism. Indeed this is one of the foundations of phylogenetic inference and the principle of common descent. While the pattern of phylogenetic conservatism is interesting by itself, it also poses a number of problems for statistical analyses. In particular, because all species are related they cannot be treated as independent observations in analyses (e.g., [129]). For example, say we would like to determine if the seed mass of plant species is correlated with the maximum height of species. Initially the urge is to calculate a correlation co-efficient or a regression. Unfortunately, given the nonindependence of the data points (due to shared ancestry) we cannot do a simple regression or correlation. Rather we must use a phylogenetically informed comparative analysis. One may find that without performing such

---

The online version of this chapter (doi: 10.1007/978-1-4614-9542-0_7) contains supplementary material, which is available to authorized users

N.G. Swenson, *Functional and Phylogenetic Ecology in R*, Use R!,
DOI 10.1007/978-1-4614-9542-0_7, © Springer Science+Business Media New York 2014

analyses, many reviewers and editors of a journal will be motivated to reject manuscripts that perform only phylogenetically uninformed trait correlation analyses due to its violation of basic statistical assumptions regarding nonindependence.

In this section we will first consider how to perform phylogenetically informed trait correlation analyses in R. We will then cover how to calculate whether your continuous trait data has "phylogenetic signal." Phylogenetic signal can be thought of as the degree to which similarity in trait values between species can be predicted upon their relatedness. If there is phylogenetic signal in the data, then phylogenetic comparative methods are necessary for robust statistical analyses of trait correlations.

7.2.1 Independent Contrasts

The phylogenetically independent contrasts (PICs) method formalized by Felsenstein [126] remains one of the most widely used approaches for quantifying the correlation between two traits while acknowledging the phylogenetic nonindependence of species. An attractive feature of PICs is that they are conceptually easy to understand with a contrast being calculated at each internal node in the phylogeny for a trait. A contrast is simply the difference in a trait between the two daughter nodes weighted by their branch lengths. Thus, if the daughter branches are equally long and the first daughter has a trait value of 120 and the second daughter of 160, then the contrast for the parent node is 40. The estimated trait value of the daughter nodes is calculated as the mean of its daughters weighted by their branch lengths such that the contrasts can be calculated recursively from the tips of the phylogeny toward the root.

Contrasts are calculated for a single trait at a time. A correlation of traits in a contrast setting requires that contrasts be calculated at each node for each trait. If the traits are correlated after accounting for phylogenetic nonindependence, it is expected that the contrast values themselves, which are now statistically independent, are correlated. A positive correlation indicates that both traits were larger in the same daughter node and smaller in the other daughter node. A negative correlation indicates trait A was larger in the first daughter lineage and smaller in the second daughter lineage, whereas trait B was smaller in the first daughter lineage and larger in the second daughter lineage. A lack of correlation indicates that the traits are not correlated after accounting for phylogenetic nonindependence. When calculating the correlation or regression between a set of contrasts for two traits, it is essential to force the correlation or regression through the origin. This is because contrasts are calculated, for example, where the second daughter node trait value is subtracted from the first, but whether this value is negative for both traits or positive for both traits the meaning is equivalent. The programming of PICs can be difficult due to the recursive nature of the calculation. Fortunately, a simple function in the *ape* package called `pic()` is available for easy use.

We begin by reading in the example trait data matrix for this chapter using the `read.table()` function assuring that the taxa names are read in as row names.

```
> traits <- read.table("comparative.traits.txt", sep =
"\t", header = T, row.names = 1)
```

Next we read in the example phylogeny for this chapter by first loading the *ape* package and the using its `read.tree()` function.

```
> library(ape)
```

```
> my.phylo <- read.tree("comparative.phylo.txt")
```

We can now perform a phylogenetically independent contrast analysis for the first trait in our trait data matrix using the `pic()` function. We make sure to sort the trait data by the order of the tip labels in our phylogeny.

```
> pic.x <- pic(traits[my.phylo$tip.label, 1], my.phylo)
```

```
> pic.x
```

We see that we have produced a vector of values. These are the contrast values for each internal node sorted from the root toward the most terminal internal node in the phylogeny. Next we calculate the contrast value for each node for the third trait in our trait data matrix again assuring that the trait data are sorted in the same order as the names in the phylogeny.

```
> pic.y <- pic(traits[my.phylo$tip.label, 3], my.phylo)
```

Next we take the node contrasts and perform a simple linear regression forcing the regression line through the origin by placing a −1 after the regression equation.

```
> pic.output <- lm(pic.y ~ pic.x - 1)
```

```
> summary(pic.output)
```

We can see that our two traits have a nonsignificant positive correlation after accounting for phylogenetic nonindependence. We can also see that a regression slope is provided by no intercept value because we forced the regression through the origin. We can now plot the contrasts and the fit regression line using the `plot()` and `abline()` functions (Fig. 7.1).

```
> plot(pic.y ~ pic.x)
```

```
> abline(pic.output)
```

We have now calculated a correlation between two traits accounting for phylogenetic nonindependence using Felsenstein's independent contrast method. The PIC method for calculating correlations is a special case of the next method we will discuss, phylogenetic generalized least squares regression, and the two will therefore produce very similar results.

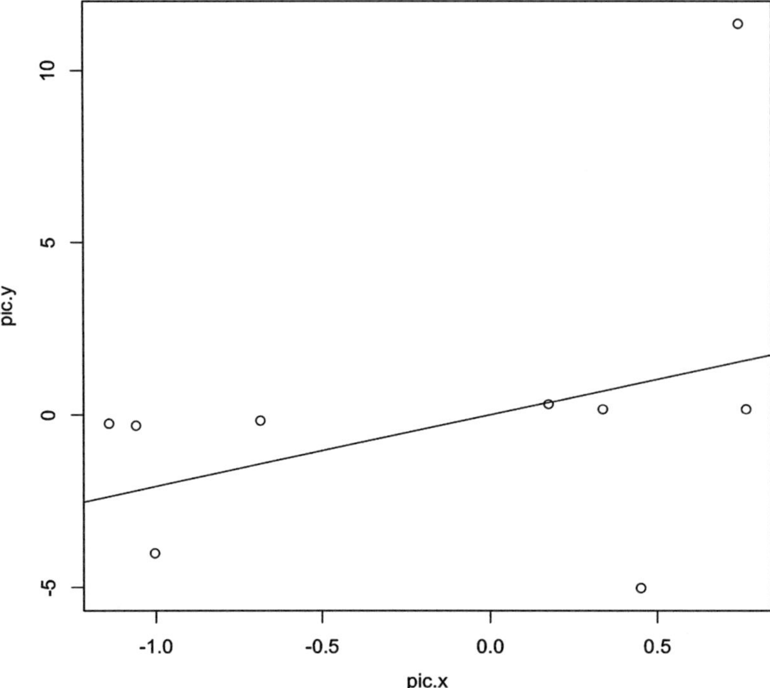

Fig. 7.1 A plot of the contrasts for trait three against the contrasts for trait one with a *line* representing our regression model forced through the origin

7.2.2 *Phylogenetic Generalized Least Squares*

An alternative method to PICs, or actually a more general method, for quantifying the correlation of two traits is to perform a phylogenetically informed regression. Phylogenetic generalized least squares (GLS) regression has been used for over a decade and provides a regression alternative that allows those more comfortable with a regression framework, though again the results from the PIC approach will be very similar [130–132]. The phylogenetic GLS approach incorporates phylogenetic nonindependence into generalized linear models in the form of a phylogenetic variance–covariance (VCV) matrix. Specifically, the regression model uses an assumed model of trait evolution to generate an expected correlation structure (i.e., nonindependence) in the data. We begin by first generating the expected correlation structure under a Brownian Motion model of trait evolution using the `corBrownian()` function in the *ape* package.

```
> cor.bm <- corBrownian( , phy = my.phylo)
```

Now that the correlation structure is defined we can calculate a simple generalized linear regression for our first two traits. First we extract the data for our first two traits sorting them in the same order as the tip labels in the phylogeny.

```
> trait.1 <- traits[my.phylo$tip.label, 1]

> trait.3 <- traits[my.phylo$tip.label, 3]
```

The GLS regression can now be calculated using the gls() function in the *nlme* package.

```
> pgls.output <- gls(trait.3 ~ trait.1, correlation =
cor.bm)
```

When running the gls() function you may receive a warning that the data are assumed to be in the same order as the names in the phylogeny. Because we have made sure to sort the data in the prior two lines of code, this is not a problem, but when performing analyses on your own datasets.

```
> summary(pgls.output)
```

In the summary for the regression model we can find the coefficients and their associate P-values to determine whether we have a significant relationship between our two traits. In this instance, again we have a nonsignificant relationship. The summary also contains a log-likelihood and AIC value that may be useful when comparing different regression equations.

A useful aspect of the phylogenetic GLS approach for quantifying the correlation between traits is that the correlation structure assumed in the model does not need to be only defined by a Brownian Motion model of trait evolution. Alternative models can be assumed or even fit to the data and this potentially makes the phylogenetic GLS approach more general and flexible than the PIC approach. In the next subsection we will discuss an approach that does not assume or fit a model of trait evolution, thereby making it more controversial, called phylogenetic eigenvector regression.

7.2.3   Phylogenetic Eigenvector Regression

The two approaches for quantifying the correlation between two traits while taking into account phylogenetic nonindependence incorporated a model of trait evolution and form their own class of comparative methods. An alternative class utilizes Euclidean distances from the phylogeny in the form of a phylogenetic distance matrix and assumes no model of trait evolution. This approach is called phylogenetic eigenvector regression. The lack of a model of trait evolution is one of the reasons why this phylogenetic eigenvector regression is so controversial.

Phylogenetic eigenvector regression was first proposed by Diniz-Filho et al. [133] and utilizes the same general class of eigenvector-based statistics designed to account for spatial autocorrelation in data with the exception that the goal is to account for phylogenetic autocorrelation. Eigenvector regression begins with the derivation of a distance matrix, whether spatial or phylogenetic, that can be used to define the autocorrelation of data points. The distance matrix is then used in a principle components analysis to derive spatial or phylogenetic eigenvectors (i.e., the "scores" derived from a PCA). The scores, or locations of spatial locations or taxa in the phylogeny on the PC axes, are then used in a multiple linear regression model as independent variables along with one trait as a dependent variable and one or more traits as additional independent variables. The question of how many PC axes to include as independent variables is often a sticking point for critics of the eigenvector approach as the number included may seem arbitrary and adding too many axes will lead to over-fitting of the regression model. A "broken stick" method is generally applied to spatial eigenvector regression, and often phylogenetic eigenvector regression, for determining the number of PC axes to utilize in the regression model. The method states that PC axes should be added until the cumulative explained variance is surpassed by that expected by a broken stick. For example, the expected cumulative variance from a broken stick for PC1, PC1+PC2, and PC1+PC2+PC3 is 50 %, 75 %, and 87.5 %. If the cumulative variance explained by the PC1, PC1+PC2, and PC1+PC2+PC3 is 66 %, 80 %, and 86 %, only the scores from the first two PC axes should be utilized as independent variables in the multiple regression model.

Because what phylogenetic eigenvectors actually describe can seem foreign and hard to visualize, it is best that we first calculate the eigenvectors for our phylogeny and then plot the scores of the first few PC axes on our phylogeny. This will aid our ability to understand what are really the independent variables in the regression model. Keep in mind that, like any PCA, the following analysis will seek to produce orthogonal axes that explain different amounts of the total variance captured in your phylogeny. The first PC axis will generally explain the majority of the variation and typically involves the most basal split in the phylogeny weighting the taxa derived from one daughter branch on one side of the first PC axis and the taxa derived from the second daughter branch on the other side of the first PC axis. Thus, as a general rule, the PC axes will generally first describe major basal splits in the phylogeny and gradually more terminal splits. To perform our phylogenetic eigenvector regression, we first generate a phylogenetic distance matrix from our phylogeny.

```
> p.dist.mat <- cophenetic(my.phylo)
```

Next we use this distance matrix in a PC analysis using the `princomp()` function.

```
> phylo.pca <- princomp(p.dist.mat)
```

We can investigate the cumulative variance explained by the PC axes by summarizing the output. This cumulative variance can be used along with the

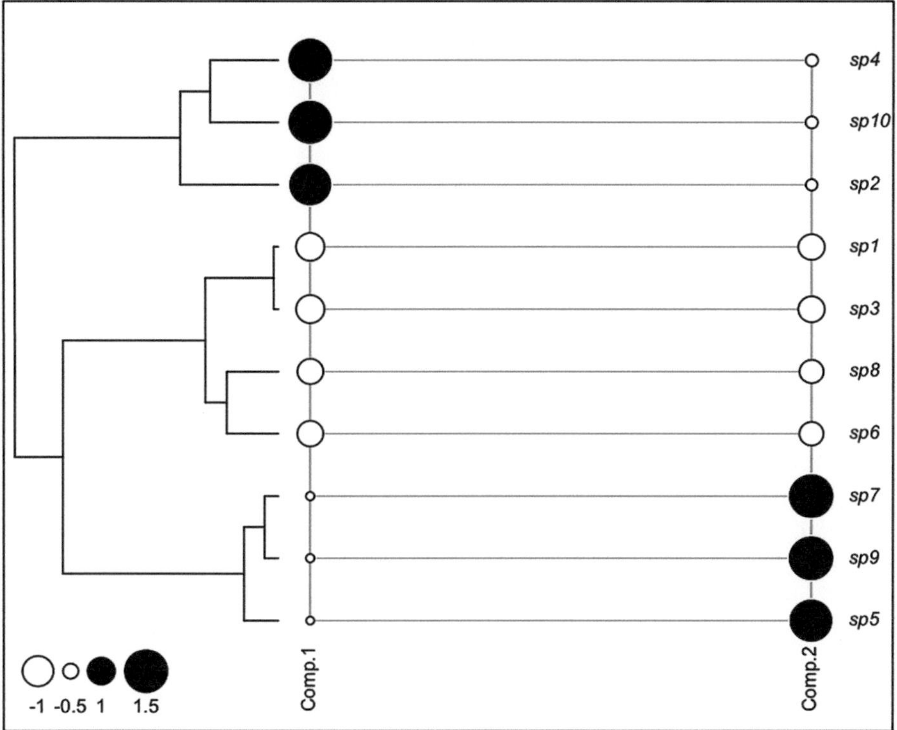

Fig. 7.2 A plot showing the phylogenetic eigenvector values for the first two eigenvectors. You will see that the first eigenvector essentially splits the species relative to the most basal node in the phylogeny and the second eigenvector begins to differentiate more terminal relationships within one clade

broken stick method to determine the number of PC axes to use in the multiple regression.

```
> summary(phylo.pca)
```

For the sake of expediency, we will not invoke the broken stick method and simply utilize the first two PC axes for our multiple regression. Instead we will simply use the first to PC axes. To visualize how the taxa in our phylogeny load on these axes, we can use the *adephylo* package.

```
> library(adephylo)
> library(phylobase)
```

First we use the phylo4d() function to merge our PC axis data with our phylogenetic tree.

```
> obj4d <- phylo4d(my.phylo, phylo.pca$scores[ ,1:2])
```

Second we can use the `table.phylo4d()` function to plot the phylogeny with the "trait" (i.e., the locations of our species on the first two PC axes) plotted to the right with the trait values indicated by open and filled circles of different sizes (Fig. 7.2).

```
> table.phylo4d(obj4d, cex.sym =.7, cex.lab = .7)
```

Now that we can visualize how the locations of species on the first two PC axes are arrayed on our phylogeny, we can now run a multiple linear regression using our second trait as the dependent variable and the scores from the first two PC axes and our first trait as the independent variables.

```
> pev.mod <- lm(traits[my.phylo$tip.label, 2] ~
traits[my.phylo$tip.label, 1] + phylo.pca$scores[ ,1] +
phylo.pca$scores[ ,2])

> summary(pev.mod)
```

The summary of the model provides the overall fit and reports whether the PC axes and our first trait are significantly related to our second trait. Because this approach can be performed rather simply in a linear model context and because spatial eigenvectors are familiar to many ecologists, phylogenetic eigenvector regression is still frequently used in ecology. However, most comparative methods researchers and evolutionary biologists have shied away from or all out rejected phylogenetic eigenvector regression. Verbal arguments against the eigenvector approach are that it does not assume or fit a model of trait evolution, the PC analysis manipulates the original structure of a phylogeny in a way that may not be biologically meaningful, and the selection of the number of eigenvectors to use in the multiple regression appears arbitrary. Simulation-based critiques of the eigenvector approach have pointed out that phylogenetic eigenvectors have far less statistical power to detect the phylogenetic signal in trait data [134]. The result of this is generally an advocacy for using phylogenetic GLS or its specialized case of PIC correlation when calculating the correlation between a set of traits while accounting for phylogenetic non-independence.

7.3 Quantifying Phylogenetic Signal

The term "phylogenetic signal" is increasingly used in the ecological literature and it is often conflated with multiple different terms and concepts. For example, phylogenetic signal is often used as a synonym for "phylogenetic conservatism" when they can be two completely different things where signal could indicate evolution of traits along a phylogeny that is less labile than expected under a Brownian Motion model of evolution (i.e., a random walk) and phylogenetic conservatism could mean no trait evolution at all (i.e., stasis in a trait through time). This creates a great deal of confusion in the literature. I highly recommend discussions by Blomberg and

Garland [135], Losos [136] and Wiens [137] to get a sense of what various terms typically mean and how they are used. Here we will generically define phylogenetic signal as the degree to which variation in species trait values is predicted by the relatedness of species. I will refrain from providing a quantitative description because, as we will soon see, there are multiple approaches for quantifying phylogenetic signal. Remember that not all approaches are equally valid or robust and I will highlight this as we discuss each metric.

7.3.1   *Mantel Test*

Our discussion of phylogenetic signal in trait data will begin with the Mantel Test. The Mantel Test quantifies the correlation between two distance matrices. In the context of quantifying phylogenetic signal, the two distance matrices are a phylogenetic distance matrix and a univariate or multivariate trait distance matrix. A Mantel Test using a univariate trait distance matrix is used to quantify the signal in a particular trait, whereas an analysis using a multivariate trait matrix may be used to quantify the signal in species function. The Mantel Test is appealing for a couple of reasons. First, it is conceptually easy to understand that a positive correlation between phylogenetic distance and trait distance may indicate phylogenetic signal in a trait and a negative correlation may indicate phylogenetic antisignal. Second, Mantel Tests can be calculated in many software packages that non-phylogeneticists (e.g., ecologists) utilize making this approach for calculating phylogenetic signal more readily available at least in the past prior to R packages for phylogenetics.

Despite the appeal of Mantel Tests for quantifying phylogenetic signal and their popularity in the past, they are now infrequently used. A good reason for this is that more sophisticated metrics that have been generated over the past decade and recent work by has shown that Mantel Tests tend to have less statistical power to detect phylogenetic signal when compared to the more recently derived metrics [138]. In particular, this work shows that Mantel Tests suffer from a lack of statistical power and inflated type I error rates compared to Blomberg's K statistic that we will discuss below. Thus, most do not recommend the use of a Mantel Test to quantify phylogenetic signal over other metrics. I therefore only present it here for completeness and historical context. Mantel Tests can be performed rather easily in R using the *vegan* package for ecological analyses. We will first install and load this package.

```
> install.packages("vegan")

> library(vegan)
```

We can now read in a phylogenetic tree into R using `read.tree()` and our example phylogeny for calculating phylogenetic signal.

```
> my.phylo <- read.tree("comparative.phylo.txt")
```

A phylogenetic distance matrix can now be generated using `cophenetic()`.

```
> p.dist.mat <- cophenetic(my.phylo)
```

We now read in the trait dataset for the phylogenetic signal example for this chapter.

```
> traits <- read.table("comparative.traits.txt",
sep = "\t", header = T, row.names = 1)
```

It is critical that the trait data be in the same order as the names in the phylogenetic distance matrix. We can assure this is the case by sorting the traits matrix using the order of the row names in the phylogenetic distance matrix.

```
> traits <- traits[row.names(p.dist.mat), ]
```

We can now make a trait distance matrix using all traits in the matrix. Note that we are assuming here that the trait data are not significantly co-varying. This is likely not the case in most datasets and a principle components analysis to reduce data redundancy may be necessary (see Chap. 4 for discussion).

```
> trait.dist.mat <- dist(traits, method = "euclidean")
```

A phylogenetic distance matrix and trait distance matrix have now been generated and we have all the information needed to conduct a Mantel Test. Remember that a positive correlation between the two matrices indicates the difference between species in trait space measured using Euclidean distance is positively related to the phylogenetic distance between those two species measured using branch lengths. In other words, a significant positive correlation is taken as evidence of phylogenetic signal in the trait data. Conversely, a significant negative correlation is taken as evidence of phylogenetic antisignal. The Mantel Test can be calculated using the `mantel()` function.

```
> mantel(p.dist.mat, trait.dist.mat)
```

We have found a significant positive correlation between phylogenetic and trait distance. In the above, I mention that this is taken as evidence of phylogenetic signal in the trait data, but I would like to reiterate again that Mantel Tests are now rarely utilized and have reduced statistical power. In the next subsection we will discuss Blomberg's *K*, which has been shown to have significantly more statistical power to detect phylogenetic signal [138].

7.3.2　Blomberg's K and Significance Tests

The next metric of phylogenetic signal that we will discuss is perhaps the most widely used metric in ecology. The metric was developed by Blomberg and colleagues [139] and seeks to quantify the degree to which variation in a trait is explained by the structure of a given phylogenetic tree. This value is then standardized by an expectation derived from Brownian Motion trait evolution on the observed phylogenetic tree. This standardization procedure is useful in that it potentially allows for the comparison of values of their statistic, which they termed K, across phylogenies and studies.

The first step in calculating the K statistic for a trait dataset is to quantify the phylogenetically corrected mean squared error for the observed trait data. We begin by extracting the trait data in the first column of our trait data matrix and sorting the data in the same order as the taxa names in our phylogeny. The trait vector is equivalent to the X parameter in the Blomberg et al. [139] equations.

```
> trait.vector <- traits[my.phylo$tip.label , 1]
```

For simplicity we will store the number of taxa in our trait dataset so that we can link this value to the parameter (n) in the equations utilized by Blomberg et al. [139].

```
> n <- length(trait.vector)
```

A phylogenetic variance–covariance (VCV) matrix is generated next. Recall, that the diagonal of this matrix is the root to tip distance for each species in the phylogeny and the off diagonal values are the distance to the most recent common ancestor between two species. Thus the expected different between two taxa is expected to scale with their off diagonal value in this matrix. The VCV matrix we are generating is equivalent to the parameter V in the Blomberg et al. [139] equations.

```
> my.vcv <- vcv.phylo(my.phylo)
```

We can now compute the inverse of the phylogenetic VCV matrix, which corresponds to the V^{-1} parameter in the Blomberg et al. equations using the `solve()` function.

```
> inv.vcv <- solve(my.vcv)
```

The next step is to calculate the phylogenetically corrected mean trait value. This value is equivalent to the estimated trait value at the root node of the phylogenetic tree [139, 140]. This value can be quantified by summing the product of the inverse VCV and trait vector and dividing this value by the sum of the inverse VCV. This resulting value is equivalent to the $\hat{a}$ parameter in the Blomberg et al. equations.

```
> root.value <- sum(inv.vcv %*% trait.vector)/
sum(inv.vcv)
```

We are now ready to quantify the observed mean squared error (MSE_0) of the trait data (X) from the phylogenetically corrected mean ($\hat{a}$). The equation from Blomberg et al. for this calculation is as follows:

$$MSE_0 = \frac{\left(X - \hat{a}\right)'\left(X - \hat{a}\right)}{n - 1}$$

This can be calculated in R by quantifying the product of a transposed matrix of trait deviations from the estimated root value and a non-transposed matrix of the same values. The value is then divided by the number of taxa minus one (i.e., the degrees of freedom).

```
> MSEo <- (t(trait.vector - root.value) %*%
(trait.vector - root.value)) / (n - 1)
```

The mean squared error (MSE) of the data given the VCV can be calculated as

$$MSE = \frac{\left(U - \hat{a}\right)'\left(U - \hat{a}\right)}{n-1}$$

where U is the VCV transformed trait vector. This calculation can be simplified in R as the product of the transposed trait deviations from the estimated root value, the inverse of the VCV and the trait deviations from the estimated root value. This value is then divided by the degrees of freedom.

```
> MSE <- (t(trait.vector - root.value) %*% inv.vcv %*%
(trait.vector - root.value)) / (n - 1)
```

The observed MSE_0 value can now be divided by the MSE value to produce an index of phylogenetic signal where larger values indicate signal and lower values indicate no signal or antisignal, but Blomberg et al. [139] point out that these values are not comparable across phylogenies and studies and therefore should be standardized. The standardization used is to divide this observed ratio by the ratio expected given a Brownian Motion model of trait evolution on the observed phylogeny. This expected ratio can be calculated as:

$$\frac{MSE_0}{MSE} = \left(\frac{1}{n-1}\right)\left(trV - \frac{n}{\Sigma\Sigma V^{-1}}\right)$$

where trV is the trace of the VCV matrix (i.e., the sum of the diagonal values). This can be easily calculated in R.

```
> expected.MSEo.MSE <- (sum(diag(my.vcv))
- (n / sum(inv.vcv))) / (n - 1)
```

We now have the observed and expected ratios of mean squared errors. We are now ready to calculate the K value, which is simply the observed ratio divided by the expected ratio.

$$K = \frac{observed\,\dfrac{MSE_0}{MSE}}{expected\,\dfrac{MSE_0}{MSE}}$$

```
> (MSEo / MSE) / expected.MSEo.MSE
```

Values of *K* of one therefore indicate that the observed variation in the trait data is predicted by the structure of the phylogenetic tree (i.e., the phylogenetic VCV) under a Brownian Motion model of trait evolution. Values of *K* greater than one indicate more phylogenetic signal than expected from Brownian Motion, whereas values less than one indicate less than expected.

There are now multiple R packages than can calculate the *K* value for a given trait and phylogeny, but we will implement the `phylosig()` function in the *phytools* package because it provides a randomization test to assess the significance of the observed *K* value.

```
> library(phytools)
```

The *K* of a trait can be calculated by providing a sorted vector of the trait data in the same order as the names on the phylogeny.

```
> phylosig(my.phylo, traits[my.phylo$tip.label, 1],
  method = "K", test = FALSE)
```

Because *K* is a descriptive statistic, it has left many wondering if their *K* value is "significant." To assess significance, a randomization test is implemented in the `phylosig()` function that simulates random trait datasets on the phylogeny to generate null distribution from which a *P*-value can be calculated. This method can be invoked by turning the test option to true and the number of simulations can also be specified.

```
> phylosig(my.phylo, traits[my.phylo$tip.label, 2],
  method = "K", test = TRUE, nsim = 1000)
```

Alternative randomization methods could potentially be implemented such as the permutation of species names along the tips of the phylogeny and recalculating the *K* value each time to provide a null distribution, but using the currently available function outlined above is likely equally or more robust.

7.3.3   *Pagel's Lambda*

The next metric we will explore also, in a sense, considers how the distribution of a trait on a phylogeny compares to that expected by Brownian Motion. The main difference here, though, is that we seek the transformation of our original phylogeny that best predicts the distribution of our traits on the phylogeny under a Brownian Motion model of trait evolution. The phylogeny is transformed using a parameter called lambda. It is therefore helpful to first understand how the phylogeny is transformed and how it appears under different levels of lambda before considering how this value is used to quantify phylogenetic signal in trait data.

The lambda transformation of a phylogeny typically utilizes values of lambda ranging somewhere between zero and one or slightly higher than one. Lambda values that are too high can generate negative branch lengths, which is impossible.

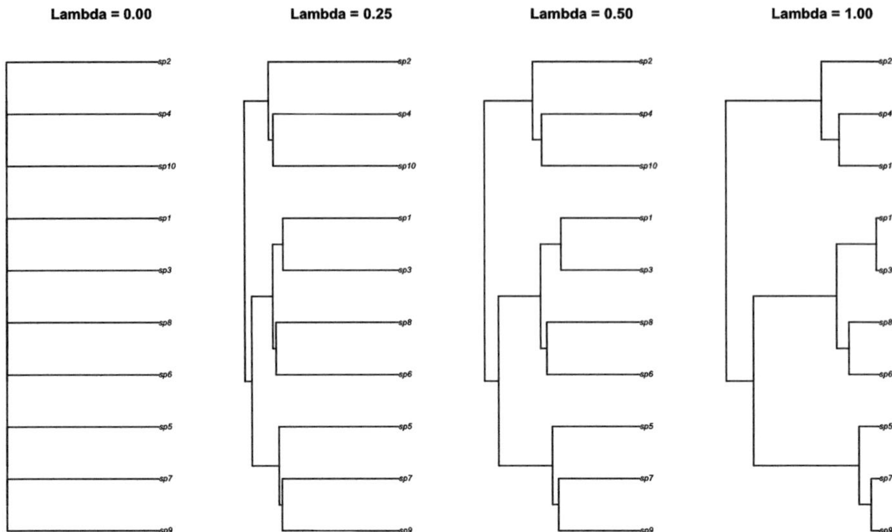

Fig. 7.3 A plot showing four lambda transformations of our original phylogeny. A lambda transformation of one retains the original structure and a transformation of zero produces a star phylogeny where all species are equally related

A lambda value of one maintains the original branch lengths of the phylogeny, whereas a lambda value of zero collapses your original phylogeny to a "star phylogeny" where all species are equally related emanating from a single soft polytomy. This may be hard to visualize in your head, so it is best to simply plot a series of lambda-transformed trees to understand what is happening. The R package *geiger* can perform a lambda transformation of a phylo object using the command `transform()`. We first load this package and perform the lambda transformation.

```
> library(geiger)

> lambda.phylo <- transform(my.phylo, "lambda")
```

The `lambda.phylo` object that we output is actually a small function that will produce a lambda transformed tree given an input value for lambda. For example, we can output a phylo object that contains a transformation of our original phylogeny using a lambda value of 0.22.

```
> lambda.phylo(0.22)
```

To visualize and compare our phylogeny transformed using different values, we can plot four versions side-by-side. I will use lambda values of 0, 0.25, 0.50, and 1.00, but you can use other values (Fig. 7.3).

```
> par(mfrow = c(1,4))
> plot(lambda.phylo(0))
> title("Lambda = 0.00")
> plot(lambda.phylo(0.25))
> title("Lambda = 0.25")
> plot(lambda.phylo(0.5))
> title("Lambda = 0.50")
> plot(lambda.phylo(1))
> title("Lambda = 1.00")
```

The far right panel displays a transformation using a lambda of one, which retains the original phylogeny. As we move toward the left with increasingly small lambda values, we see that the terminal branches become longer and the internal nodes of the phylogeny are pushed basally until we reach a lambda of zero giving a star phylogeny where all species are equally related. As the internal nodes are pushed more basal in the phylogeny, we expect that sister lineages will have more divergent trait values under a Brownian Motion model of trait evolution. The reason for this is that the time since their divergence or distance to their most recent common ancestor (MRCA) is getting larger and given that the expected variance in a trait scales with the branch length under Brownian Motion we can expect that the longer the path length to the MRCA, the greater the expected divergence in traits between two lineages. With this in mind and a general understanding of how changes in lambda influence the structure of the phylogeny, we can now discuss how lambda may be used as a metric of phylogenetic signal in trait data.

The general idea for using lambda to measure phylogenetic signal is to search for the lambda value that transforms the original phylogeny such that the observed distribution of traits on the tips of the phylogeny is mirrored by that expected under Brownian Motion on the transformed phylogeny [141]. A very low lambda increases the distance between sister taxa and therefore their expected trait difference under Brownian Motion. Thus, a lambda-based test of phylogenetic signal that results in low lambda values indicates very little phylogenetic signal in the trait data given the original phylogeny and a high lambda value indicates relatively more phylogenetic signal in the trait data given the original tree. Indeed, if the value of lambda resulting from the analyses is one, then our original phylogeny and a Brownian Motion model of trait evolution best match our observed distribution of traits on the tips of the original phylogeny. The lambda measure of phylogenetic signal in trait data can be calculated using the *phytools* package and the same `phylosig()` function we used to calculate Blomberg's *K*. In this example, we will again compute the phylogenetic signal in our trait in the second column of the example trait data matrix.

```
> phylosig(my.phylo, traits[my.phylo$tip.label, 1],
  method = "lambda", test = FALSE)
```

The function outputs the lambda value and the log-likelihood. Given that it is not possible to search all possible values of lambda for transforming our phylogeny to fit the data, an optimization must be used to search and estimate the value. Thus, the result output above likely does not match yours using the same data. To see this you can run the above line of code several times. The output lambda values will be slightly different each time reflecting minor differences in the optimization output. A log-likelihood for the reported lambda is also output by the `phylosig()` function. This value can be compared to the log-likelihood of lambda equal to zero for a significance test using a likelihood ratio test. Changing the test flag to true from the default of false results in a likelihood ratio test being computed.

```
> phylosig(my.phylo, traits[my.phylo$tip.label, 1],
method = "lambda" test = TRUE)
```

The function now has output the additional information of the log-likelihood of lambda equal to zero and the *P*-value from the likelihood ratio test. Note that the test compares your reported lambda to the null hypothesis of a lambda equal to zero or a "star phylogeny." It is therefore a true null hypothesis where relatedness explains none of the trait similarity between species. This is somewhat different from the randomization tests we used for Blomberg's *K* above where the null was a distribution of *K* values from simulated data where the null hypothesis is not necessarily a star phylogeny with a *K* of zero.

7.3.4   Standardized Contrast Variance, Unstandardized Contrast Means, and Randomization Tests

The next two approaches for quantifying the phylogenetic signal in trait data utilize the phylogenetic independent contrasts for a single trait. Recall, that the contrast for a node describes the magnitude of the difference of the trait values for daughter nodes. This difference can be standardized weighting by branch lengths or unstandardized where all branch lengths are set to one. Thus, large contrast values indicate that daughters are very divergent in their traits, which may indicate the lack of phylogenetic signal. The mean contrast value has therefore been used in the past to quantify the phylogenetic signal in trait data (e.g., [33]). This mean is typically calculated using the unstandardized contrast values. We therefore must set all branch lengths in our phylogeny to one prior to calculating the independent contrasts. First we make a duplicate copy of our original phylogeny.

```
> my.phylo.new <- my.phylo
```

Next we use the `compute.brlen()` function in the *ape* package to change all branch lengths to one. If you wish to check that the branch lengths have indeed changed, you can check the distribution of edge lengths in both the original and new phylo objects.

```
> my.phylo.new <- compute.brlen(my.phylo.new, 1)
```

We can now calculate the contrast values for one of the traits in our trait matrix and take a mean value. We will assure that the order of the trait values in the vector align with the names of taxa on the phylogeny. We will utilize the trait values in the second column of our trait data matrix. Also remember that your phylogeny must be fully bifurcating.

```
> mean(pic(traits[my.phylo.new$tip.label, 1],
my.phylo.new))
```

The output value is the mean contrast for our trait across all internal nodes in our phylogeny with branch lengths set to one. Large values indicate that one average daughter nodes have large trait differences, whereas small values indicate little difference. Though, we have no reference to what is "large" and what is "small" and cannot state whether the observed value is higher or smaller than that expected. Thus, we cannot infer phylogenetic antisignal or phylogenetic signal, respectively. Thus, we need a null distribution that we can compare our observed value against. This can be accomplished using a permutation test where taxa names are shuffled on the tips of the phylogeny. This can be accomplished by producing a small function, which we can replicate. In particular, we will write a function that randomizes the names of taxa on the phylogeny and runs the `pic()` function.

```
> pic.shuffle.mean <- function(x){

    ## Shuffle the tip labels on the phylo.
    x$tip.label <- sample(x$tip.label,

    length(x$tip.label), replace = FALSE)

    ## Calculate mean PIC with shuffled names.
    mean(pic(traits[x$tip.label, 2], x))
}
```

This function can now be replicated 999 times to produce a null distribution of mean contrast values using the `replicate()` function.

```
> replicate(999, pic.shuffle.mean(my.phylo.new))
```

We can now compare where the observed value lands in this null distribution of mean contrast values. We can accomplish this by combining the observed value with the 999 random values and calculating the rank of the first (i.e., the observed) value in the distribution.

```
> rank(c(mean(pic(traits[my.phylo.new$tip.label,1],
my.phylo.new)), replicate(999,
pic.shuffle.mean(my.phylo.new)))))[1]
```

We would like to perform a two-tailed test for phylogenetic signal and antisignal. If the output rank (a.k.a. quantile) value is low that indicates our mean observed contrasts is generally lower than most random values. This would be indicative of significant phylogenetic signal if the output rank value was less than or equal to 25 (e.g., $25/1{,}000=0.025$; $P=0.025$). Conversely, if the rank value is greater than or equal to 975 (e.g., $975/1{,}000=0.975$; $P=1-0.975=0.025$), then this indicates significant phylogenetic antisignal because the average contrast value is significantly higher than expected by random.

An alternative to quantifying the average unstandardized contrast value as an indicator of phylogenetic signal is to calculate the variance of the standardized contrasts [135, 139]. The rational for this approach is that if there is a small degree of variation in contrast values across internal nodes in the phylogeny, this should result in closely related species being more similar in trait values due to the nested structure of the phylogeny. Conversely, if closely related species are not similar, the variance in contrast values across the phylogeny will be elevated. This metric is generally calculated using a phylogeny with branch lengths and therefore standardized contrasts. The variance in standardize contrasts can be simply calculated for our example trait two as:

```
> var(pic(traits[my.phylo$tip.label, 1], my.phylo))
```

The randomization test for the variance approach takes the same format except that we must slightly change the function that we will replicate to calculate a variance and not a mean.

```
> pic.shuffle.var <- function(x){
      ## Shuffle the tip labels on the phylo.
      x$tip.label <- sample(x$tip.label,

      length(x$tip.label), replace = FALSE)
      ## Calculate variance of the PICs with shuffled
      ## names.
      var(pic(traits[x$tip.label, 2], x))
  }
```

This function can now be replicated 999 times to produce a null distribution of PIC variance values and where the observed variance lands in that distribution using the `replicate()` and `rank()` functions.

```
> rank(c(mean(pic(traits[my.phylo$tip.label,2],
my.phylo)), replicate(999, pic.shuffle.var
(my.phylo)))))[1]
```

Similar to the mean PIC approach above, rank values less than or equal to 25 are indicative of phylogenetic signal and values greater than or equal to 975 are indicative of phylogenetic antisignal. It is important to remember that in both instances the significance of the rank values is determined by the number of replicates plus one. Thus, if we replicated the analyses with 9,999 random trees, significant phylogenetic signal would be found if the rank was less than or equal to 250.

The analysis of mean and variance of the unstandardized and standardized contrast values, respectively, will often produce different results. Of the two approaches the variance approach is currently implemented to a greater degree due to the fact it takes advantage of branch length information. It should be stated though that generally these two approaches are not used with near the frequency that the above K and lambda approaches are utilized.

7.3.5   *Phylogenetic Eigenvectors*

The final method we will discuss for measuring phylogenetic signal is phylogenetic eigenvectors. Similar to the Mantel Test, phylogenetic eigenvectors have been shown to have a reduced statistical power to detect phylogenetic signal in trait data [134]. This issue and the degree to which selecting the number of eigenvectors to incorporate into the regression model is or is not arbitrary are major sticking points that have resulted in the infrequent use of phylogenetic eigenvectors for measuring phylogenetic signal. We therefore cover their calculation primarily for historical context and completeness rather than advocating for their use.

As before we begin by generating the phylogenetic eigenvectors by performing a PCA on a phylogenetic distance matrix.

```
> phylo.pca <- princomp(cophenetic(my.phylo))
```

Next, we examine the summary of the output object. From this output we can apply the broken stick method (see Sect. 7.3.3) to select the number of axes to be used for our downstream regression analyses. In this example, for expediency we will simply select the first three axes and not implement the broken stick method for selecting the number of eigenvectors. To extract the eigenvectors for the first three axes for all species, we can do the following:

```
> pevs <- phylo.pca$scores[ , 1:3]
```

Now we perform a linear regression regressing the trait data in column two of our trait data matrix on the phylogenetic eigenvectors. Make sure to sort the trait data according to the order of the row names in the phylogenetic eigenvector output.

```
> summary(lm(traits[row.names(pevs), 1 ] ~ pevs))
```

The resulting summary of the linear model reports the adjusted R-squared. This value and the significant level could be used to infer the phylogenetic signal in trait data, but this approach is often criticized and may have very limited statistical power (e.g., [134]). The coefficients from this multiple regression could be used to determine which eigenvectors best explain the trait data. Please refer to Sect. 7.3.3 to see how to visualize these eigenvectors.

7.4   Quantifying the Timing and Magnitude of Trait Divergences

The above section dealt with how to quantify the phylogenetic signal in trait data, but the result from each analysis is a phylogeny-wide measure regarding the degree of signal. However, we are often interested node-level signal or antisignal and how this signal changes with the position of the node in the phylogeny. In words, we are often interested in the size of the trait divergence between two daughter nodes and whether large or small divergences tend to be correlated with the depth in the phylogeny (e.g., time). If large divergences happened relative deep in the phylogeny and terminal nodes have little divergence, this may be evidence that the group rapidly filled trait space early and began to pack this space through time. We will discuss a couple of conceptually related approaches to quantifying the timing and magnitude of trait divergence on a phylogenetic tree.

We will begin by taking the simple approach of asking whether the contrast for a trait at each node is unusually large or small. Recall, that in Sect. 7.4 we quantified the mean and the variance of contrast values in our phylogeny for a single trait. We then utilized a randomization procedure to calculate a series of random means and variances for 999 phylogenies that had their names shuffled. This code can easily be adapted to quantify the contrast at each node for the observed and random phylogenies by removing the mean calculation. Specifically, we can generate the following shuffling contrast function.

```r
> pic.shuffle <- function(x){

    ## Shuffle the tip labels on the phylo.
    x$tip.label <- sample(x$tip.label,
    length(x$tip.label), replace = FALSE)

    ## Calculate mean PIC with shuffled names.
    pic(traits[x$tip.label, 1], x)
}
```

This function can be replicated 999 times to produce a matrix with the number of rows equaling the number of nodes in the phylogeny and 999 columns using the `replicate()` function.

```r
> node.null <- replicate(999, pic.shuffle(my.phylo))
```

We can now bind our observed contrasts to this matrix as follows:

```r
> combined.obs.null
<- cbind(pic(traits[my.phylo$tip.label, 1], my.phylo),
node.null)
```

The result is a matrix with the first column being the observed contrast values and the next 999 columns being the random contrast values for those nodes. We can calculate the rank of where the observed lands in the null distribution using an apply() function with a MARGIN = 1 taking only the first column from the output given we want the rank of the observed value and not the ranks of the random values. Given that we have 999 randomizations, a rank value of less than or equal to 25 is evidence that the contrast for that node is small than expected, whereas a value greater than or equal to 975 is evidence that the contrast for that node is larger than expected.

```r
> apply(combined.obs.null, MARGIN = 1, rank)[1, ]
```

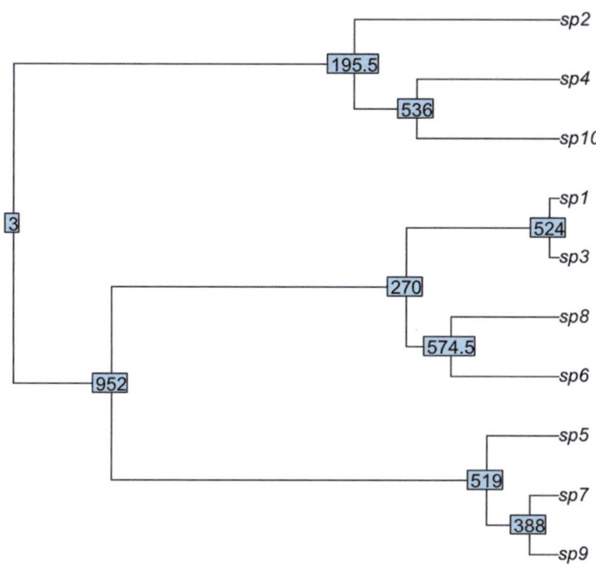

Fig. 7.4 A plot of our phylogeny where the internal nodes are labeled with the observed contrast value ranks in the distribution of 999 random contrast values we calculated. Values above or equal to 975 would indicate a higher than expected divergence or contrast and values below or equal to 25 would indicate a lower than expected divergence or contrast

The order of the resulting rank values is in the same order as the nodes in your phylo object. If you have your node labels, you can combine the ranks with the node label with a `cbind()`. If you wish to plot the ranks on the corresponding nodes in the phylogeny, you can utilize the following code (Fig. 7.4).

```
> my.ranks <- apply(combined.obs.null, MARGIN = 1,
rank)[1, ]

> plot(my.phylo)

> nodelabels(my.ranks)
```

If we wish to plot the distribution of rank values through time, we first extract the relative ages of each internal node in the phylogeny using the `branching.times()` function in the *ape* package.

```
> times <- branching.times(my.phylo)
```

The branching times produced begin at the root and end at the most terminal internal node. They are therefore in the same order as the ranks in our output. The branching times are also quantified such that the largest value is assigned to the root. For example, if the root is 50 million years old and the youngest internal node is 10 million years old, their branching times will be 50 and 10, respectively.

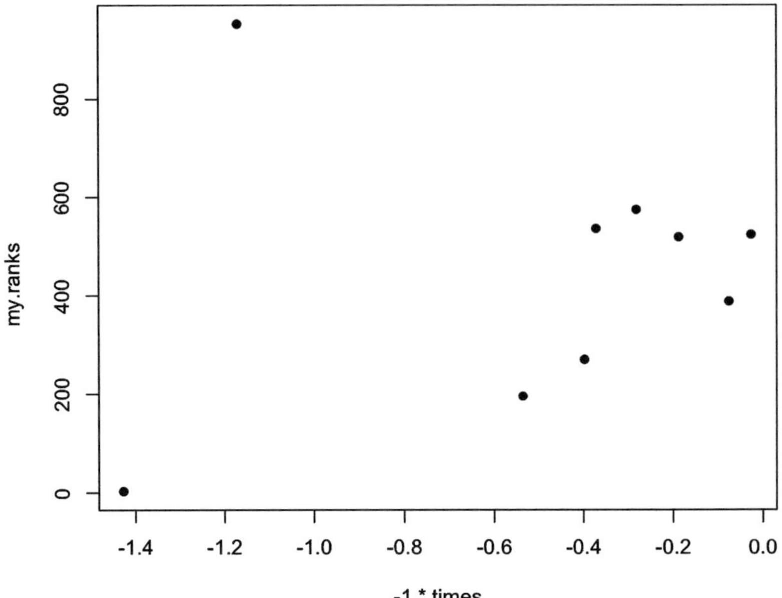

Fig. 7.5 A plot of the observed contrast value for a node ranks in the distribution of 999 random contrast values we calculated plotted against the divergence times for that node with the most basal divergence being plotted on the left side of the *x*-axis

Thus, if we make a simple plot of the data, the position of the nodes on the *x*-axis will be the reverse of their position on the *x*-axis when plotting the phylogeny. To correct this we can simply multiply the branching times by negative one such that a node that is 50 million years old is now −50 on the *x*-axis (Fig. 7.5).

```
> plot(-1 * times, my.ranks, pch = 16)
```

This procedure is useful for not only plotting ranks from contrast randomizations, but generally for plotting node-level calculations as a function of time or relative time.

The above examines contrasts at each node in the tree and compares them to a null distribution generated by shuffling of names on the phylogeny. Contrasts and name shuffling are not the only approach for quantifying the timing and magnitude of trait divergence. An alternative approach proposed by Harmon et al. [142] call Disparity Through Time (DTT) quantifies the disparity of daughter node trait values and compares the observed values to a null distribution generated by simulating trait evolution on the phylogeny many times under a Brownian Motion model. For continuous traits the disparity can be calculated using two methods. The first is the average squared Euclidean distance between the two daughter values and the second is the average Manhattan distance between the two daughter values. It is recommended

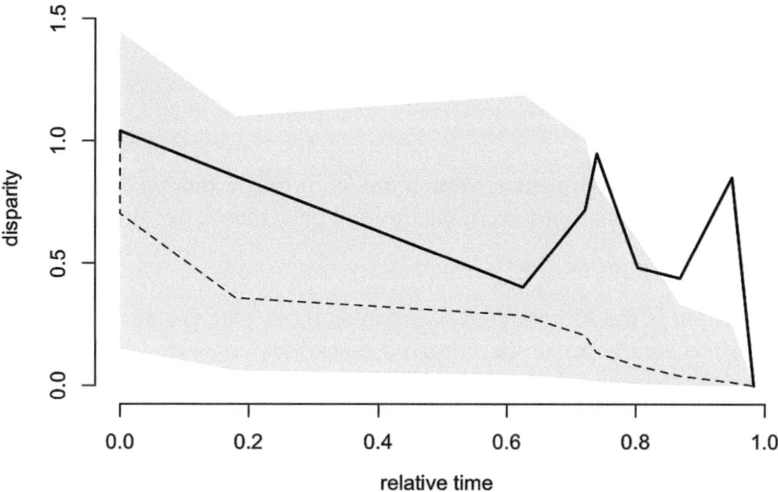

Fig. 7.6 A plot of the observed multivariate trait disparities at a node (*black line*), mean of the null distribution of disparities from a Brownian Motion model of trait evolution (*dashed line*), and 95 % envelope from the null distribution (*shaded gray*). The most basal node is on the left side of the *x*-axis. Nodes that have trait disparities or divergences that are significantly higher than that expected under a Brownian Motion model are indicated by places where the *black line* is above the *shaded area*

that Euclidean distances should be used if all trait axes being analyzed are in the same units (i.e., z-standardized traits or PCA axes) and that Manhattan distances should be used if the traits are all in different units. Given that we have focused in previous chapters on reducing co-variation in trait data matrices through the use of PC analyses, we will utilize Euclidean distances in this example. The calculation of DTT can be performed using the *geiger* package and the `dtt()` function. We will use all traits simultaneously to quantify a multivariate disparity for each node (Fig. 7.6).

```
> my.dtt <- dtt(my.phylo, traits, index ="avg.sq",
nsim = 999)
```

There are multiple outputs from the `dtt()` function and we will cover them one by one. First we ask for the names of what is contained in the output object.

```
> names(my.dtt)
```

The first output are the disparities calculated for each node ordered from the root toward the tips.

```
> my.dtt$dtt
```

The second output are the branching times for the nodes ordered from the roots toward the tips. In the `dtt()` function the branching times are transformed by subtracting root age from each value and dividing by the root age. The values therefore range from zero to one.

```
> my.dtt$times
```

The third output is a matrix holding the disparities for each node from the simulated datasets using a Brownian Motion model of trait evolution. There is one row per node ordered from root toward the tips and one column per simulation.

```
> my.dtt$sim[1:4,1:4]
```

The last output is the Morphological Disparity Index (MDI) statistic. This statistic calculates the area between the observed disparities connected by a line and the median of the expected disparities derived from the Brownian Motion simulations connected by a line. Observed disparities that are larger than the median add to the area (i.e., the MDI) and observed disparities lower than the median subtract from the area (i.e., the MDI). Thus, a positive MDI indicates the observed disparity values are one average larger than that expected from a Brownian Motion model of trait evolution, whereas negative values indicate the observed disparities are on average smaller than that expected. Negative values of MDI can be taken as evidence of an early burst in the evolution of morphological diversity followed by little morphological diversification within more terminal subclades. Conversely a positive MDI value can be taken evidence of a constant or accelerating rate of morphological diversification through time.

One result that the `dtt()` function does not directly calculate is whether individual nodes diverge significantly from the null distribution for that node derived from the Brownian Motion simulations. This is perhaps surprising given that all of the necessary information to calculate node-level significance has been output by the function. Specifically, we know the observed disparities and we have a null distribution of disparities for each node. All that is left to do is to calculate the rank of the observed value in the null distribution. This can be accomplished using the following code.

```
> apply(cbind(my.dtt$dtt, my.dtt$sim), MARGIN = 1,
rank)[1, ]
```

In this example we have once again generated 999 random values such that we have a total of 1,000 values in the distribution including the observed values. Thus, values less than or equal to 25 indicate a disparity value for a node that is lower than that expected given Brownian Motion and values greater than or equal to 975 indicate a disparity value for a node that is greater than that expected. We can plot these ranks on the phylogenetic tree using the same approach we used previously in this subsection to plot the ranks of the observed contrast values.

As you can see, the contrast and disparity approaches are conceptually very similar. The main difference is that the disparity approach provides a summary statistic of how the observed disparities generally deviate from the expectation, though it isn't too different from the use of contrasts to quantify phylogenetic signal (see Sect. 7.4), and the use of a model of trait evolution to produce the null distribution in the disparity approach rather than using the shuffling of names across the tips of the phylogeny. For this second reason the disparity approach may be preferred because it utilizes an explicit model of trait evolution and allows for the possibility of other models of trait evolution to be utilized to set the expectation.

7.5 Conclusions

In this chapter we have considered the issue of phylogenetic nonindependence between species in comparative datasets. We began by covering how to incorporate phylogenetic information into trait correlation analyses. This was followed by a demonstration of a variety of methods for quantifying the degree to which trait similarity is explained by the phylogenetic relatedness of species and how this phylogenetic signal can be calculated phylogeny wide and at individual nodes. Although there is a more extensive text in the UseR! series regarding comparative methods in R [15], and I encourage you to also consult that text, I have included this chapter in this book as measurements of trait correlations and phylogenetic signal are critical for most phylogenetic and functional diversity analyses in ecology.

It is my hope that this chapter will help ecologists to begin or continue or adventure into the world of phylogenetic comparative methods, but you will quickly find that the phylogenetic comparative methods literature is vast and full of very strong opinions on various topics. I hope that this does not discourage you in your pursuit of using phylogenetic information in your ecological analyses because at the very least it is the statistically appropriate thing to do and moreover it likely will make your analyses much more informative. The code above therefore is simply a starting point for your use of R for phylogenetic comparative methods.

7.6 Exercises

1. Simulate three trait datasets on your example phylogeny using the `fastBM()` function in the *phytools* package.
2. Calculate a correlation of the independent contrast values for two of the traits and perform a phylogenetic generalized linear model analysis with these same two traits arbitrarily selecting one as the dependent variable.
3. Calculate the phylogenetic signal in each simulated trait dataset using Blomberg's K, Pagel's Lambda, and the variance in independent contrasts.

Chapter 8
Partitioning the Phylogenetic, Functional, Environmental, and Spatial Components of Community Diversity

8.1 Objectives

The objectives of this chapter are to understand and implement methods for partitioning the functional and phylogenetic dimensions of diversity within and between communities and to quantify the relationships between traits, phylogenetic relatedness, space, and the environment simultaneously.

8.2 Background

The broadening of the concept of biodiversity to include dimensions such as phylogenetic and functional diversity has resulted in a large and ever-growing literature. The majority the research published in this realm has generally quantified the phylogenetic or functional diversity in a system and often correlates these results with some gradient or gradients. Analyses that simultaneously consider phylogenetic and functional diversity are less common, but can be found in the literature. Much more rare are studies that directly integrate phylogenetic and functional diversity metrics and/or partitioning variation in phylogenetic or functional diversity into the components that can be explained by spatial and environmental gradients and their co-variation. Thus, in many ways ecology has generally not yet successfully integrated phylogenetic, functional, environmental, and spatial information into a unified analytical framework. This limits our ability to simultaneously address in one study system the fundamental questions of what aspects of organismal function are selected by the environment to determine the spatial distribution of species and how have these important aspects of function evolved? In other words, ecologists are still by and large struggling to explicitly quantify the evolutionary fingerprint on the

The online version of this chapter (doi: 10.1007/978-1-4614-9542-0_8) contains supplementary material, which is available to authorized users

N.G. Swenson, *Functional and Phylogenetic Ecology in R*, Use R!, 173
DOI 10.1007/978-1-4614-9542-0_8, © Springer Science+Business Media New York 2014

patterns of diversity and abundance in present day ecological communities. While I am not convinced that a perfect analytical framework exists yet to address this issue, particularly in communities composed of multiple distantly related lineages (i.e., not a community or trophic level composed of a single mono-phyletic group), there are a few approaches that are available to begin to integrate phylogenetic, functional, environmental, and spatial data that are fundamental to ecological analyses and inferences. I will cover a few of these approaches in this chapter starting with the partitioning of variation in phylogenetic or functional alpha or beta diversity. After that we will examine a fairly new modification incorporating phylogenetic information into an older statistical approach to link trait data with spatial gradients in the environment. Both of the general approaches to be covered have been under-utilized in the literature in my opinion. In introducing them to you I hope they will become more widely used and that they may inspire some to generate their own novel approaches that integrate these pieces of information in a powerful and intuitive way.

8.3 Partitioning Variation in Community Functional Alpha Diversity by the Environment, Space, and the Community Phylogenetic Alpha Diversity

The functional diversity (FD) within a community or sample can be explained by several potentially interacting factors. Frequently it is assumed that the abiotic environment plays a key role in determining the FD within an assemblage particularly on larger spatial scales. However, abiotic gradients are generally spatially autocorrelated such that trend in FD along an abiotic gradient may simply be due to spatially structured processes such as dispersal limitation that are independent of the environment. Thus, we may be interested in partitioning the FD quantified within assemblages in a study system into the portion that can be explained by the abiotic gradient(s) the assemblages are found in independent of spatial distance (i.e., a pure environmental component), the portion that can be explained simply by the spatial distance independent of environmental distance (i.e., a pure spatial component), the portion that can be explained by the interaction of space and the environment or spatial autocorrelation in the environment (i.e., the space–environment interaction component), and the portion that cannot be explained by the spatial or environmental components and their interactions (i.e., the residual component). A final twist on this framework is that we may also be interested in partitioning the variance in FD that is explained by the phylogenetic diversity (PD) in the communities and how PD interacts with the spatial and environmental gradients in the system. In the following subsections we will explore how to perform such variance partitioning of alpha FD using multiple regression on distance matrices and spatial relationships defined as the Euclidean distance between communities. We follow this with a similar multiple regression-based framework, but with spatial relationships between plots being quantified using Principal Coordinates of Neighbor Matrices (PCNM) analyses.

8.3.1 *Partitioning FD Using Multiple Regression on Distance Matrices*

The goal of this subsection is to first calculate the functional diversity (FD) of the communities in our community data matrix. We will then partition the FD between communities based on the spatial and environmental distances that separate them. Then, we will augment these analyses by adding a phylogenetic component to our variance partitioning. We will use multiple regression on distance matrices (MRM), but the general concept of variance partitioning could be used to use other approaches that do not use MRM.

We start by loading the *picante* package so that we may calculate FD and reading in our community data matrix and trait data matrix.

```
> library(picante)

> my.sample <-
read.table("partition.example.sample.txt", sep = "\t",
header = T, row.names = 1)

> my.traits <-
read.table("partition.example.traits.txt", sep = "\t",
row.names = 1, header = T)
```

The following analyses all use spatial and environmental data for each community. The spatial data will be read into R in the form of a matrix with the first column being the x coordinates and the second column being the y coordinates. The row names are the community names. The environmental data contain three variables in the three columns. It is critical that these tables have the communities in the same order as that found in the community data matrix. Thus, we will sort both tables to assure this is the case.

```
> my.xys <- read.table("xy.data.txt", sep = "\t",
header = T, row.names = 1)

> my.xys <- my.xys[row.names(my.sample), ]

> my.env <- read.table("env.data.txt", sep = "\t",
header = T, row.names = 1)
```

We now can calculate a measure of FD for each of our communities. Any measure can be utilized so long as you format the output as a vector. For simplicity we will calculate the unweighted pairwise functional distance between all species in our communities. We will calculate functional distance as the Euclidean distance between the species in our system using all traits simultaneously. The assumption here is that all traits are orthogonal. If this is not the case, a PC analysis should be used first to reduce trait redundancy (see Chap. 4). We can use the mpd() function in the *picante* package to calculate the pairwise functional distance between the species in each community.

```
> MPD.output.traits <- mpd(my.sample,
as.matrix(dist(my.traits, method = "euclidean")),
abundance.weighted = F)
```

The MRM analyses to follow require a distance matrix for each of the independent variables and the dependent variable. The dependent variable in this case is the FD. Though, our FD output is a vector sorted in the same order as the rows in our community data matrix and our spatial and environmental data matrices. We would like to partition the variation in the difference in FD between communities. As such we will generate a FD distance matrix from our vector of FD values using the `dist()` function.

```
> MPD.output.t.dist <- dist(MPD.output.traits, method =
"euclidean")
```

We are now ready to calculate a multiple regression using our FD distance matrix regressed onto a distance matrix of our geographic coordinates and environmental distance. The spatial and environmental distance matrices will be generated using the `dist()` function. The MRM analysis can be using the linear model function `lm()`, but if you would like a permutation test, you could use the `MRM()` function in the *ecodist* package.

```
> library(ecodist)

> MRM(MPD.output.t.dist ~ dist(my.xys) + dist(my.env))
```

We see that our two variables combined explain 70.78 % of the variation in community FD in our study system, but this is the multiple R^2 value and not the adjusted R^2 value we need for variance partitioning. Thus we will use a simple `lm()` function.

```
> summary(lm(MPD.output.t.dist ~ dist(my.xys) +
dist(my.env)))
```

We find that the linear model explains ~62.43 % of the variation. The next goal is to partition the amount of variation in FD between the communities in our system explained purely by the environmental distance between the communities, purely by the spatial distance, by the interaction between environmental and spatial distance (i.e., spatial autocorrelation in the environment), and the variance that cannot be explained. To begin this process, we first recalculate the full regression model and extract the R^2 value from the output.

```
> MRM.Space.Env <- lm(MPD.output.t.dist ~ dist(my.xys)
+ dist(my.env))

> MRM.Space.Env.R2 <-
summary(MRM.Space.Env)$adj.r.squared
```

Then we calculate model including only the spatial distance matrix or only the environmental distance matrix and extract the R^2 values from the output.

```
> MRM.Space <- lm(MPD.output.t.dist ~ dist(my.xys))

> MRM.Space.SandE.R2 <-
summary(MRM.Space)$adj.r.squared

> MRM.Env <- lm(MPD.output.t.dist ~ dist(my.env))

> MRM.Env.SandE.R2 <- summary(MRM.Env)$adj.r.squared
```

We now have the variation explained by the full model and from the two nested models. We find that the spatial component which includes pure space and the space–environment interaction (i.e., `MRM.Space.SandE.R2`) explains ~35.42 % of the variance and we find that the environmental component which includes pure environment and the space–environment interaction (i.e., `MRM.Env.SandE.R2`) explains ~56.17 % of the variance.

The variance explained by the pure components and their interaction and the residual unexplained variation can now be calculated. The residual variation is simply one minus the variation explained by the full model.

```
> MRM.residual.R2 <- 1 - MRM.Space.Env.R2
```

The variation explained purely by the environmental component is equal to the difference between the variation explained in the full model and the variation explained by the nested model of FD regressed onto spatial distance.

```
> MRM.Env.only.R2 <- MRM.Space.Env.R2  -
MRM.Space.SandE.R2
```

The variation explained purely by the spatial component is equal to the difference between the variation explained in the full model and the variation explained by the nested model of FD regressed onto environmental distance.

```
> MRM.Space.only.R2 <- MRM.Space.Env.R2  -
MRM.Env.SandE.R2
```

Lastly, the variation in FD that is explained by the interaction between the spatial and environmental components is the variation explained by the full model minus the sum of the variation explained by the purely spatial and purely environmental components.

```
> MRM.SandE.R2 <- MRM.Space.Env.R2 - (MRM.Env.only.R2 +
MRM.Space.only.R2)
```

We have now established how the variation in FD in a series of communities or samples in a system can be partitioned into pure spatial and environmental components, their interaction, and residual variation using MRM.

8.3.2   Partitioning FD Using Principal Coordinates of Neighbor Matrices (PCNM) and Forward Selection

The above approach utilized the Euclidean distance in space separating our communities. Thus the spatial data in these analyses are interpreted on this single linear axis and there is no scale dependence in the spatial data. In other words, there is very little spatial structure in the data when using this approach. Given that most of the biological processes we are interested in presently and elsewhere have spatial structure (i.e., biological processes are scale dependent), it may be useful to model the spatial structure in our data prior to performing partitioning analyses. A frequently utilized method in ecology for modeling this structure is to use Principal Coordinates of Neighbor Matrices (PCNM), which is a special case of Moran's eigenvector maps [19]. A PCNM analysis requires a Euclidean distance matrix for the communities or samples. This matrix is then truncated by a set value, where all values less than the truncation parameter are retained and all values larger than the truncation parameter are given an unrealistically large value. This introduces a neighborhood structure that can be input into a Principal Component (PC) analysis that will produce eigenvectors that are orthogonal and structure the spatial data at different scales, where the first eigenvectors represent coarse scale spatial structure and the latter eigenvectors represent finer scale spatial structure. Some of these eigenvectors can be utilized in a multiple regression and variance partitioning. Indeed we can first select which eigenvectors best relate to our dependent variable prior to performing a multiple regression and variance partitioning. This helps us understand the scale of the spatial structure that is most closely related to our dependent variable, which itself is potentially biologically informative.

We begin by calculating the PCNM values for our spatial data by using the `pcnm()` function in the *vegan* package. We will provide the function with a Euclidean distance matrix of our geographic coordinates and a threshold value. Choosing a threshold value is an important step. It is recommended that the threshold be as low as possible to include multiple samples under the threshold (i.e., each sample or community will have at least one other sample or community in its neighborhood). When the sample or community data are collected at regular distances in one-, two-, or three-dimensional space, determining this threshold is fairly easy. When the data are spaced irregularly (i.e., spatial coordinates of plots in a continental-scale forest inventory dataset), determining the threshold value is more difficult. It may be necessary to set a large threshold to include at least one neighbor in the neighborhood radius of each sample and this runs the risk of losing information pertaining to the fine-scale spatial structure in the dataset. An imperfect solution to this problem suggested by Bocard et al. [19] is to first randomly add geographic coordinates to your dataset to artificially allow lower thresholds that can capture finer scale structure. After the PCNM has been performed, these artificial samples or communities are removed from the downstream analyses. The reason why this is an imperfect solution is that the removal of these artificial samples may mean that the eigenvectors are no longer orthogonal and this must be remembered if this approach is taken. For the sake of simplicity and expediency, in this example we will use a threshold of two. Though please keep in mind that selecting a threshold value should be done with care in real analyses.

```
> my.pcnms <- pcnm(dist(my.xys), threshold = 2)
```

In order to decide what PCNM vectors to include in our analysis we will use "forward selection." Forward selection is implemented in the *stats* package. It can be installed and loaded using the following code.

```
> install.packages("packfor", repos = "http://R-
Forge.R-Project.org")

> library(packfor)
```

We can now perform forward selection using the `forward.sel()` function and the vectors produced by the PCNM analysis and the FD values for our communities.

```
> mpd.output.traits <- mpd(my.sample,
as.matrix(dist(traits)))

> forward.sel(mpd.output.traits, my.pcnms$vectors)
```

We will extract the first and second vectors for expediency our original PCNM output and place them in a new object for regression analysis.

```
> selected.pcnms <- my.pcnms$vectors[ , c(1,2)]
```

We can now perform the same series of `lm()` functions we used in the previous subsection, but substituting the PCNMs selected instead of the raw *xy* distance. One could also add phylogenetic diversity as an independent variable to this regression model to also partition the PD component of the observe FD [82].

We have now seen how to partition the alpha FD in a community into spatial, environmental, and phylogenetic components and their interactions. In the next section we will partition phylogenetic and functional beta diversity into purely spatial and environmental components and their interaction.

8.4   Variance Partitioning of Phylogenetic or Functional Beta Diversity Along Environmental and Spatial Gradients

The previous section discussed the partitioning of phylogenetic and functional alpha diversity. Partitioning of alpha diversity is far less common than the partitioning of beta diversity. The partitioning of phylogenetic and functional beta diversity is the focus of this section where we ask whether the compositional dissimilarity between pairs of communities in our study system is best explained by their spatial or environmental distance or the co-variation in spatial and environmental gradients in the system.

8.4.1   *Beta Diversity and Multiple Regression on Distance Matrices*

A multiple regression on distance matrices (MRM) approach can be useful for relating phylogenetic or functional beta diversity to spatial and environmental distance particularly if one does not wish to model any scale-dependent spatial structure in the data. The MRM approach is a simple extension of the partial Mantel test, which has been a popular approach in the beta diversity literature. We will not cover the partial Mantel approach because the MRM approach is now preferred, but a partial Mantel analysis can be performed using the `partial.mantel()` function in the *vegan* package. The approach for using MRM for partitioning beta diversity is nearly identical to that used above to partition alpha diversity gradients. The only difference is that we use the output distance matrix from a phylogenetic or functional beta diversity analysis, such as D_{pw}, as the dependent variable. Therefore we will use very similar code and example calculating phylogenetic beta diversity where I will only point out the differences. We will use the same phylogenetic, trait, spatial, and environmental data that we used in the preceding section. To begin we will calculate the unweighted pairwise phylogenetic beta diversity, D_{pw}, for our example system using the `comdist()` function in the *picante* package.

```
> Dpw.output <- comdist(my.sample, cophenetic(my.
phylo), abundance.weighted = F)

> my.beta.xys <- read.table("xy.beta.data.txt", sep =
"\t", row.names = 1, header = T)
```

The output distance matrix representing the D_{pw} between all samples or communities is now used as the dependent variable for the full and nested MRM models that follow.

```
> summary(lm(Dpw.output ~ dist(my.beta.xys) +
dist(my.env)))

> MRM.Space.Env <- summary(lm(Dpw.output ~
dist(my.beta.xys) + dist(my.env)))$adj.r.squared

> MRM.residual.R2 <- 1 - MRM.Space.Env.R2

> MRM.Space <- summary(lm(Dpw.output ~
dist(my.beta.xys)))$adj.r.squared

> MRM.Env <- summary(lm(Dpw.output ~
dist(my.env)))$adj.r.squared

> MRM.Pure.Space <- MRM.Space.Env - MRM.Env

> MRM.Pure.Env <- MRM.Space.Env - MRM.Space

> MRM.Joint.S.E <- MRM.Space.Env - MRM.Pure.Space -
MRM.Pure.Env
```

As you can see, the code is essentially the same as that we used for the alpha diversity partitioning in the previous section. In theory, one could partition the functional beta diversity in your study system using the spatial and environmental distance matrices and the phylogenetic beta diversity distance matrix.

8.4.2   Partitioning Beta Diversity Using Principal Coordinates of Neighbor Matrices (PCNM) and Forward Selection

As we have discussed in the previous section on partitioning alpha functional diversity, it is often useful to model the spatial structure of your data because biological processes have scale-dependent spatial structure. The modeling of this spatial structure can be accomplished by using Principal Coordinate analysis of Neighbor Matrices (PCNM). In Sect. 8.3.2 we cover PCNM in more detail and the reader is advised to quickly read that text prior to proceeding with the following code.

The partitioning of beta diversity using a PCNM approach is nearly identical to that we used for alpha diversity above with the exception that we will perform a PC analysis on the output beta diversity matrix. In this case we will use the phylogenetic beta diversity output from Sect. 8.4.1. We first perform a PCNM analysis on our spatial distance matrix using a threshold value of two. Choosing the threshold value is important and we are simply using this value for expediency. For a discussion of how to determine the threshold value for your dataset, see Sect. 8.3.2 or the discussion in Bocard et al. (CITE).

```
> my.pcnms <- pcnm(dist(my.xys), threshold = 2)
```

We now deviate slightly from the PCNM approach we used for alpha diversity. Here we perform a PC analysis of the beta diversity matrix to produce the orthogonal eigenvectors that will become the dependent variables in our regression analysis.

```
> Dpw.pcoa <- pcoa(Dpw.output)
```

The phylogenetic beta diversity eigenvectors are now used in a forward selection analysis to determine which PCNM eigenvectors are the best predictors of our beta diversity eigenvectors. Recall, that the first eigenvectors generally represent coarse-scale spatial structure and later eigenvectors generally represent fine-scale spatial structure. Thus, the spatial eigenvectors that are selected by this process provide us some information regarding the scale dependency and spatial structure of the beta diversity in our system.

```
> forward.sel(Dpw.pcoa$vectors, my.pcnms$vectors)
```

The selected eigenvectors can now be used in a series of linear models as we have done in the Sects. 8.3.1 and 8.4.1 to partition the variation in beta diversity while acknowledging scale dependency in spatial structure via a PCNM approach.

```
> adonis(Dpw.output ~ my.pcmns$vectors[ , c(1,2)] *
  my.env, permutations = 999, method = "euclidean")
```

8.5   Integrating Phylogenetic, Trait, Environmental and Spatial Information to Quantify the Role of Abiotic Filtering During Community Assembly

To this point we have integrated phylogenetic and trait diversity with environmental and spatial data to partition the diversity. This approach is typically used in the context of testing niche versus neutral community assembly mechanisms. A related, but alternative to diversity partitioning, approach is to integrate phylogenetic, trait, environmental, and spatial information to ask what traits are filtered into communities (i.e., that show lower than expected diversity), what lineages are also filtered into these communities, and what are the environmental and spatial contexts of this filtering? In other words we would like to know how traits are filtered into communities, what is the phylogenetic imprint on the traits, and how is filtering distributed across spatial and environmental gradients. A few approaches have been proposed to address these fundamental questions. One is the trait–habitat–clade (THC) approach of Mayfield et al. [143], but this approach suffers from utilizing discrete environmental and trait data. The requirement of discrete habitat types in particular is a serious limitation for most spatial analyses. We therefore focus on a proposed approach that is more flexible both in allowing for mixed (i.e., continuous and categorical) traits and continuous spatial data.

The approach we will cover here is an extension of the **RLQ** approach. The original **RLQ** approach utilizes ordination of a **R** matrix containing environmental and spatial data, a **Q** matrix containing trait data for the species, and a **L** matrix that is a community data matrix that can be used to link the **R** and **Q** matrices. This approach traditionally been used to quantify the relationship between traits and the environment across a series of samples or communities. Recently, Pavoine et al. [144] extended the **RLQ** approach to incorporate phylogenetic information. In this extended approach, the **L** matrix is still the community data matrix and is subjected to a Correspondence Analysis [145]. The **R** matrix has rows representing the samples or communities with columns that are the environmental data and spatial variables. The environmental data can be a community by environmental parameter value and the spatial variables data can be the coordinates of the communities in two columns or a neighborhood matrix. Both the environmental and spatial matrices are subjected to a Principal Components Analysis (PCA). The **Q** matrix now has species in the system as rows with raw traits or a trait distance matrix subjected to a PCA or Principal Coordinates Analysis (PCoA), respectively, and a phylogenetic distance matrix subjected to a PCoA. The modified **R** and **Q** matrices are then linked using the **L** matrix as is done a tradition **RLQ** analysis. In the following we will generate the matrices and then utilize the source code provided by Pavoine et al. [144] for performing and plotting the modified **RLQ** analysis.

We start by loading the *ade4* package in R and reading in our environmental matrix.

```
> library(ade4)

> my.env <- read.table("env.data.txt", sep = "\t",
header = T)
```

Next we read in our example Newick phylogeny using the `read.tree()` function in the *ape* package. This produces an object of class "phylo."

```
> my.phylo <- read.tree("partition.example.phylo.txt")
```

The phylo class object we now have will need to be converted to a "phylog" class object. The phylog class is used by the *ade4* package for phylogenetic trees. It contains not only a Newick tree in the object but also a great deal of additional information including a phylogenetic distance matrix. We will use the `newick2phylog()` function to convert our phylogeny. This function requires the text of a Newick phylogeny, which we can obtain by using the `write.tree()` function.

```
> my.phylog <- newick2phylog(write.tree(my.phylo))
```

We now perform a Correspondence Analysis of our community data matrix to generate our **L** matrix.

```
> COA.sample <- dudi.coa(my.sample, scan = FALSE, nf =
dim(my.sample)[2] - 1)
```

Next we perform a PCA on our raw coordinate data weighting by the rows of the **L** matrix. Recall, that we can utilize a neighborhood matrix instead. An example neighborhood matrix could be a binary matrix where communities in a grid that abut are scored as one and communities that do not abut are scored as zero. An alternative would be to generate a neighbor matrix as we did for the PCNM analyses in the previous sections using a threshold value. The output from this analysis is one half of the **R** matrix.

```
> PCA.xy <- dudi.pca(my.xys, COA.sample$lw, scan =
FALSE, nf = dim(my.xys)[1] - 1)
```

The environmental data forms the second half of the **R** matrix. Prior to subject the environmental data matrix to a PCA, we will standardize the data by dividing the logged values by the difference between the maximum and minimum value for that column.

```
> stand.value <- apply(my.env, MARGIN = 2, max) -
apply(my.env, MARGIN = 2, min)

> stand.envs <- sweep(my.env, 2, stand.value, "/")
```

The logged and standardized environmental data can now be used in a PCA weighted by the rows of the **L** matrix.

```
> PCA.env <- dudi.pca(stand.envs, COA.sample$lw, scale
= FALSE, scan = FALSE, nf = dim(stand.envs)[2])
```

A PCoA can now be performed on our trait data weighting by the **L** matrix. We will assume that our trait data are orthogonal and all are related to the environmental gradients of interest. In Pavoine et al. [144], a formal analysis of what traits to include in this calculation is presented where a regular **RLQ** analysis (i.e., `rlq()` in the *ade4* package) is performed with a permutation of trait values to determine which traits are significantly related to the environmental variables. A distance matrix is then calculated only from these traits and input into the PCoA.

```
> PCO.traits <- dudi.pco(dist(traits, method =
"euclidean"), COA.sample$cw, full = TRUE)
```

The output forms one half of the **Q** matrix. The other half of the **Q** matrix is generated by performing a PCoA on the phylogenetic distance matrix weighting by the columns of the **L** matrix. Our phylog object contains the phylogenetic distance matrix as `Wdist`.

```
> PCO.phylo <- dudi.pco( as.dist( as.matrix(
my.phylog$Wdist)[ names(my.sample), names(my.sample)
]), COA.sample$cw, full = TRUE)
```

The output of the PCoA on our phylogenetic distance matrix is the second half of the **Q** matrix. We are now ready to run the phylogenetic modification of the **RLQ** analysis proposed by Pavoine et al. [144]. This analysis not provided in a currently R package so we will provide R with the source code from Appendix 5 of the Pavoine et al. [144] paper using the `source()` function.

```
> source("Pavoine.app5.txt")
```

This source code contains the `rlqESLTP()` function that will perform the modified **RLQ** approach given the matrices we have generated above. The name of the function gives the order of the matrices to provided to the function.

```
> rlq.output <- rlqESLTP(PCA.env, PCA.xy, COA.sample,
PCO.traits, PCO.phylo, scan = FALSE, nf =2)
```

We can now calculate the proportion of the covariance of the phylogeny and traits (**Q** matrix) with space and environment (**R** matrix) for each eigenvector produced by the analysis. In this example, proportion of the covariance explained by the first axis can be calculated as follows:

```
> rlq.output$eig[1] / sum(rlq.output$eig)
```

We see that the first axis explains ~82.37 % of the variation. If we wish to visualize this axis in space, we simply use the `plot.rlqESLTP()` function provided in the source code that we read into R. This can be accomplished by providing our output object, the *x* and *y* coordinates of our communities, telling the function which axis we want to plot (`ax = 1`) and that we want to plot the axis in space (`wh = "S"`) (Fig. 8.1).

```
> plot.rlqESLTP(rlq.output, xy = my.xys, ax = 1, wh =
"S")
```

If we wish to visualize this axis on the phylogeny we plot the same function by providing our output object, our phylogeny, telling the function which axis we want to plot (`ax = 1`) and that we want to plot the axis in relation to the phylogeny (`wh = "P"`) (Fig. 8.2).

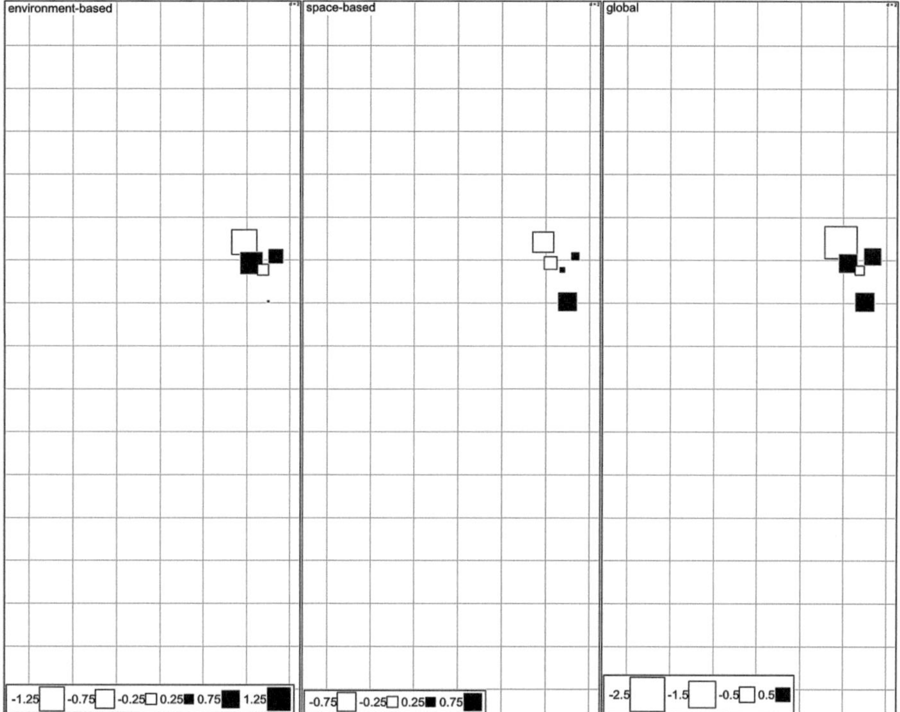

Fig. 8.1 A plot of the **RLQ** output where we plot the distribution of the first axis in space

```
> plot.rlqESLTP (rlq.output, phy = my.phylog, ax = 1,
wh = "P")
```

From these results we can see that the resulting first axis underlies ~82.37 % of the co-variation of the phylogeny and traits with space and environment. We see that this is co-variation has a spatial structure from the west to east in our study system and that this spatial structure is generally randomly distributed across our phylogeny. Thus we are able to determine how traits are arrayed along spatial and environmental gradients and on the phylogeny to link the evolutionary history of traits that are filtered into communities with spatial and environmental gradients.

8.6   Conclusions

The integration of phylogenetic and functional diversity individually into ecology is still a relatively new phenomenon. The simultaneously integration of both of these components along with spatial and environmental information is an incredibly new phenomenon that will likely become commonplace in the coming decade. Thus, although the literature and R code on this topic is sparser than other topics

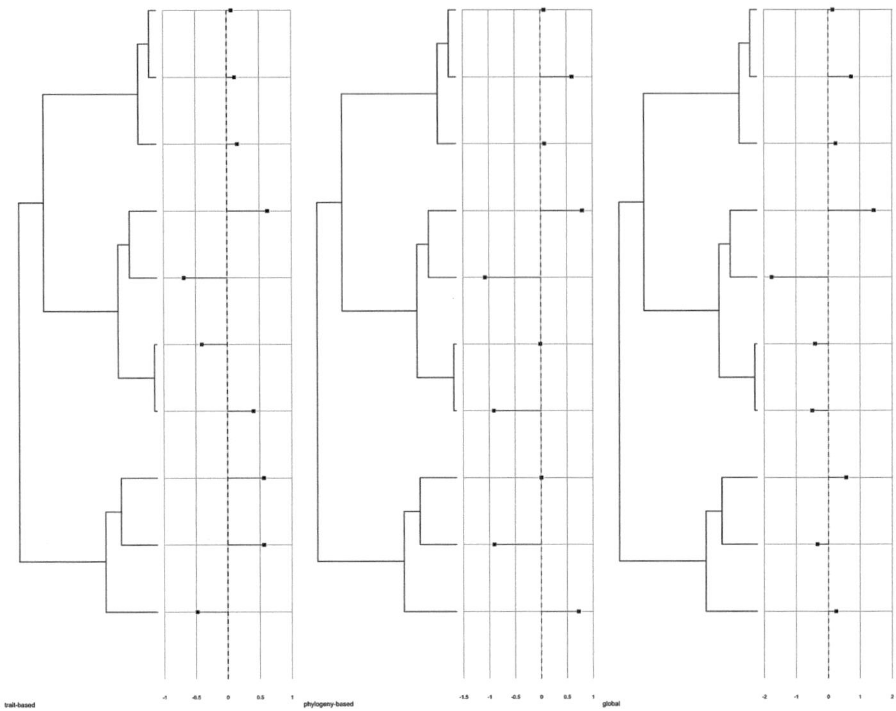

Fig. 8.2 A plot of the **RLQ** output where we plot the distribution of the first axis on the phylogeny

covered in this book, I felt it was important to cover this important emerging topic. In particular, we covered how the variation in the observed phylogenetic alpha and beta diversity in a system can be partitioned into pure spatial and environmental components and their interaction. We followed this by not only doing the same for functional diversity but also adding a phylogenetic component into the list of independent variables explaining the variation in functional alpha and beta diversity. We finished with an exciting and relatively new analytical framework proposed by Pavoine et al. [144] that extends the "RLQ approach" to incorporate phylogenetic information to address how the environment filters lineages with particular traits and how this process plays out over space in the study system.

It is my expectation that variance partitioning analyses, as we performed in the beginning of the chapter, will continue to be important in ecology and perhaps become even more important with their implementation in R and use in a phylogenetic and functional diversity context. Fortunately, the general conceptual approach for these analyses will likely not change radically in the near future. The RLQ-style integration of phylogenies and traits with environmental and spatial gradients will likely see many new approaches proposed in the near future that use a similar or a very different statistical framework, but the conceptual approach will likely not be much different given the question being addressed is fundamental to ecology. Thus, I hope that the description of the general approaches in this chapter allows the reader to utilize these same R-based analyses in their research, but to also understand the approaches that will emerge in the literature in the coming years.

8.7   Exercises

1. Quantify the Jaccard's Index for your example dataset using the `vegdist()` function in *vegan* and partition the variance explained by space and the environment and their joint effect.
2. Repeat the analyses in number 1 above, but also include a measure of phylogenetic beta diversity as an explanatory variable and its joint effects with space and the environment.
3. Quantify the functional beta diversity for your example dataset. Next, partition the variation in the functional beta diversity by spatial and environmental gradients, the Jaccard's Index values, and the interactions between these three variables.

Chapter 9
Integrating R with Other Phylogenetic and Functional Trait Analytical Software

9.1 Objectives

The objectives of this chapter are to quickly cover how to use R to interface with analytical software that is written in other programming languages. The chapter will focus primarily on integrating R commands with the software Phylocom written in C [14]. I have chosen to focus on this program because it carries out many of the types of analyses covered in this book. Although I have focused on this particular program, the general principles of how to call other programs from R and integrate them into your R-based analyses remain the same.

9.2 Background: The Development of Eco-Informatics Tools for Phylogenetic- and Functional Trait-Based Ecology

In other places in this book, we have discussed the long history of performing phylogenetic and functional diversity analyses in ecology in some cases dating back nearly a century. Despite this long history, only recently has the ecological literature seen a dramatic increase in phylogenetically and functionally based analyses. Although conceptual advances can explain some of this, I would argue that the majority of this increase is due to the development of freely available analytical and eco-informatics tools. Indeed over the course of approximately 5 years, community ecology went from a discipline that rarely generated phylogenetic trees and had few analytical tools to analyze those trees where available to one where phylogenetic trees, though often crude, could be estimated for entire communities within seconds or minutes and analyzed with powerful software. I think it would be hard to underestimate the impact of this development on ecology in general and plant community ecology in particular.

Recently developed analytical and eco-informatics tools have had a large role in the increasing the number of phylogenetically and, to a lesser extent, functionally

N.G. Swenson, *Functional and Phylogenetic Ecology in R*, Use R!,
DOI 10.1007/978-1-4614-9542-0_9, © Springer Science+Business Media New York 2014

based analyses in ecology and the most widely utilized of these tools are coded in C. Although these C programs are open source, ecologists who are not familiar with C and perhaps more fluent in R have found it difficult to utilize or modify the analyses available in the C programs. For example, the two most widely used C programs in phylogenetic community ecology are Phylocom and Phylomatic [14, 56]. Ecologists not accustomed to using command lines have often found these programs difficult to understand or those ecologists not fluent in C have found it difficult to change the source code in these programs to run additional analyses that are not a "canned" function. Thus, when R packages such as *picante* were derived, many phylogenetic community ecology analyses became unlocked to the large number of R using ecologists. One issue though that these users have run into is that the original C program is faster for null model analyses for a series of reasons of which may be the use of `for()` loops in R. For this reason I frequently encounter researchers who say an analysis is "impossible" in R using the available packages when I know it is easily possible in the original C programs. One solution to this issue is writing faster R code, but an alternative option is to use R to communicate with the C program. This approach may increase speed, but it also has the benefit of perhaps allowing you to run analyses that you cannot code (yet) in R. Further, it may allow you to pull out additional information regarding your analyses that the original C program may not allow. For example, the C program Phylocom runs a series of null models and for phylogenetic alpha diversity metrics will produce a standardized effect size value and *P*-value and for phylogenetic beta diversity will only produce a standardized effect size value. In both cases, it is not easy or possible for the user to see the null distribution and in the case of phylogenetic beta diversity it isn't possible to obtain a *P*-value.

In the following sections I will describe how to use R to communicate with the Phylocom program written in C. The goal here will be to simply understand how to call another program from R. After we have mastered this task and seen how to run many phylogenetic and functional diversity analyses in Phylocom while in R, I will discuss how to use R to "unlock" parts of Phylocom to produce all null model analyses and how to seamlessly integrate your R-based analyses with Phylocom.

9.3 Phylocom

Many attribute the rise in phylogenetic analyses of communities to a paper written by Cam Webb in 2000 in The American Naturalist regarding the phylogenetic structure of tree assemblages in Borneo [28]. While this paper was very interesting and did provide the original formulations of the Net Relatedness Index and Nearest Taxon Index, the article in and of itself was not the reason for the tidal wave of phylogenetic community ecology that arrived in the subsequent decade. Rather, it was the eco-informatics tools developed by Webb and colleagues [28] to generate and analyze phylogenetic trees for the species in plant communities that were made freely available to the research community. The tools they developed were and are available in the C programs Phylocom and Phylomatic. In this section we will discuss how to integrate Phylocom and R and we will cover Phylomatic in the next section. Phylocom is the main analytical software produced by Webb and colleagues to quantify the phylogenetic diversity of communities and to perform comparative methods such as independent contrasts and measures of phylogenetic signal. This program is still frequently utilized by researchers, but has been forgotten or avoided by researchers comfortable only in R. My goal here is to simply demonstrate how this powerful program can be accessed and utilized by calling it from R. We will begin with quantifying phylogenetic and functional diversity metrics with null models, follow this with quantifying independent contrasts and phylogenetic signal, and end with a demonstration of how to randomize data in R and provide that data to Phylocom to produce null distributions. Before we do any of this though, you will need to download and unzip the latest version of Phylocom from https://www. phylodiversity.net/phylocom.

9.3.1   Quantifying Phylogenetic and Functional Diversity and Dispersion in Phylocom

Now that we have downloaded Phylocom onto our computer and unzipped the directory, we are ready to begin interfacing it with R. First, for simplicity, let us set our working directory in R to the same directory on our computer that holds the Phylocom program. For Mac users this is the "mac" directory in the Phylocom directory. For PC users this is the "w32" directory in the Phylocom directory. Note that the PC version is 32 bit, which can cause memory issues for large datasets. I am a Mac user so I will set my working directory as follows:

```
> setwd("/Users/swenson/phylocom-4.2/mac")
```

We will now write some example files to these directories that we can use for analysis. We will use the example Phylocom phylogeny, community data, and trait files from the *picante* package.

```
> library(picante)
```

```
> data(phylocom)
```

First, we extract the example phylogeny and write it to our working directory in a Newick file format using `write.tree()`.

```
> my.phylo <- phylocom$phylo
```

```
> write.tree(my.phylo, "my.phylo.txt")
```

Second, we extract the example community data matrix and write it to our working directory in the Phylocom format where each row is a species in a particular community with the first column being the community name, the second column being the abundance of that species in that community, and the third column being the species name. It is critical that this file is sorted by the first column, that a species does not occupy more than one row per community, and that the species names perfectly match the names of the species in the phylogeny. These first two requirements are handled by the `writesample()` function in *picante* and the third requirement is not a concern here given we are using an example dataset, but should be checked by you or using the `treedata()` function in *geiger*.

```
> my.sample <- phylocom$sample

> writesample(my.sample, "my.sample.txt")
```

Last, we extract the example trait data matrix and write it to our working directory. Again it is critical that the names in the trait data match those in the phylogeny and community data. The first row in Phylocom trait data matrices contains information regarding the type of trait data (e.g., continuous, categorical, etc.). In this instance our trait data is continuous so this row will include a "3" as the type with the word "type" in the first column. The second row provides the name of each trait in the dataset with the word "name" in the first column. The remaining rows in the trait data matrix contain species names in the first column and trait values in the remaining columns. This format can be produced by using the `writetraits()` function in *picante*.

```
> my.traits <- phylocom$traits

> writetraits(my.traits, "my.traits.txt")
```

If you now look in your working directory you will see these files have been written to your hard drive.

```
> list.files(getwd())
```

You can open these files with a text editor to examine their format and to quickly see if there are any obvious errors that may have occurred. If the files look fine, then we are ready to proceed with some simple analyses using Phylocom from R.

The following commands will differ slightly depending on whether you are using a Mac operating system or Windows. As a Mac user I will be using the `system()` function, but if you are a Windows/PC user, you can simply replace `system()` with the `shell()` function. These functions are used to call a program that can be run in your terminal or command line. For example, we can first use R and `system()` to call Phylocom.

```
> system("./phylocom")
```

The description of the Phylocom software is printed to my R console. Here I can see all of the functions available in this program and how to call them, but first a couple of quick points about how I used the `system()` function. First you will note that I used quotes around the program that I am calling. So if you had a program called SuperProgram and the executable was in your working directory, you would replace "./phylocom" with "./SuperProgram." Second you will notice that I had a "./" in front of the program name. This is because I have not placed Phylocom in my path on my Mac and you likely have not so that the program is not universally accessibly by your computer. Thus, unless we set the program into our path (see Phylocom user manual) we can simply use "./" prior to the program name for Mac OSX and Unix operating systems. For Windows users the "./" is not required.

Now that we can call the Phylocom program from R, let us first calculate the Faith's Index to discover how to use Phylocom. Like many command line programs, Phylocom works by first telling the command line the program you want to use (i.e., phylocom), then the command or method you would like to use in that program, and finally perhaps some options may be selected. The program we want to use is phylocom and the method we would like to use is "pd" since that is the command for calculating Faith's Index in Phylocom. We follow these two commands by telling the program the name of our phylogeny file using the "-f" switch and the name of our community data file using the "-s" switch.

```
> system("./phylocom pd -f my.phylo.txt -s
my.sample.txt")
```

The program will look for your phylogeny and community data files in your working directory, so please make sure they are there. You should see a matrix print on your R console with five columns and seven rows. The first row gives the column names and each row is a community in your community data matrix. The species richness in each community is given in the "ntaxa" column, the Faith's Index is given in the "PD" column, the summed branch lengths for the entire phylogeny is given in the "treeBL" column, and the Faith's Index represented as a proportion of the summed branch lengths of the entire tree is given in the "propTreeBL" column. If you wanted to actually save this output and not have it only printed to your R console, you could tell Phylocom to write a .txt file to your working directory with the output using a ">" followed by the file name you want to assign the output.

```
> system("./phylocom pd -f my.phylo.txt -s
my.sample.txt > my.faiths.output.txt")
```

You should now see that file in your working directory and you can read it into R using `read.table()` if you wanted to do further analyses or plotting. Next we can calculate the mean pairwise phylogenetic distance and the mean nearest phylogenetic neighbor and perform a null modeling analysis simultaneously in Phylocom using the "comstruct" command. We will again provide the information regarding where to find the phylogeny and community data files, but we will also use the "-m" switch to inform the program what null model to utilize and the "-r" switch to tell the program how many randomizations to run.

```
> system("./phylocom comstruct -f my.phylo.txt -s
my.sample.txt -m 0 -r 99")
```

The output should print out on your R console, but if you wanted you could save it to your hard drive if you used a ">" followed by the file name. We only ran 99 randomizations but this could easily be switched to 999 or higher. We used "-m 0" to indicate we wanted to use null model type "0" which shuffles the names of species on the phylogeny. An independent swap of the community data matrix null model could be invoked by using "-m 3" instead. If we examine the output we see again, there is one row per community in our dataset. The species richness is the "ntaxa" column, the mean pairwise phylogenetic distance is the "MPD" column, the mean of the null distribution of MPD values is in the "MPD.rnd" column, and the standard deviation of that null distribution is in the "MPD.sd" column. The next column is named NRI for Net Relatedness Index. Though here we should pause to consider a small difference between Phylocom and R output that is of major importance. First recall that the NRI is a standardized effect size (S.E.S.) for the MPD metric. In the *picante* R package in the code and the equations we used in Chap. 6, we defined that the S.E.S. is the observed value minus the mean of the null distribution and that value divided by the standard deviation. Now if we consider the output from Phylocom we see that the NRI value is *not* the (MPD − MPD.rnd)/MPD.sd. Rather it is this value multiplied by negative one. Therefore in Phylocom when the observed MPD is larger than the mean of the null distribution, the NRI value is negative. In R when the observed MPD is larger than the mean of the null distribution, the NRI is positive. The multiplication by negative one in Phylocom is due to this calculation being used in the original work by Webb [28] that introduced the NRI and NTI metrics. Thus, it is critical to know whether a researcher has calculated his or her NRI or NTI value using Phylocom or the *picante* package because a negative value in one program has the opposite meaning in the other. It is not completely uncommon for me to see this issue come up in the papers of colleagues where the exact opposite inference has been made. So please be cautious. The best method is to simply look at the observed value and the mean of the null distribution and then consider the S.E.S. value. By doing this you can see whether a S.E.S. value is being multiplied by negative one or not. Now that we have discussed that issue we can return to the results on our screen. The next two columns indicate where the observed value lands in the null distribution. The first reports how many null values are greater and the second column reports how many null values are lower. These rank values can be used to estimate *P*-values. The next six columns are in a similar format except they concern the mean nearest taxon distance and the Nearest Taxon Index (NTI; again multiplied by negative one). The last column reports the number of randomizations used in the null model. This number and the type of null model are also reported in the very first row of the output.

The NRI and NTI analyses above only considered the species as present or absent, but the metrics can be weighted using abundance by using the "-a" switch.

```
> system("./phylocom comstruct -f my.phylo.txt -s
my.sample.txt -m 0 -r 99 -a")
```

Lastly, we can output all of the individual random MPD and MNTD by adding a "-v" to the end of this line of code. This permits the user to examine the null distributions themselves.

Now we are ready to compute phylogenetic beta diversity using Phylocom via R. Specifically, we can calculate the phylogenetic D_{pw}, D_{pw}', D_{nn}, and D_{nn}' metrics. We will start by not running any null model analyses and use the D_{pw} metric, which is calculated using the "comdist" command in Phylocom.

```
> system("./phylocom comdist -f my.phylo.txt -s
my.sample.txt")
```

The result is a matrix where the community names are on the rows and columns and the values are the pairwise phylogenetic dissimilarity between each set of communities. The diagonal values correspond to the MPD values. We can again weight this D_{pw} calculation by abundance to produce the D_{pw}' values by adding the "-a" switch.

```
> system("./phylocom comdist -f my.phylo.txt -s
my.sample.txt -a")
```

The D_{nn} and D_{nn}' values for our communities can be calculated using the "icomdist" function in Phylocom.

```
> system("./phylocom comdistnt -f my.phylo.txt -s
my.sample.txt")

> system("./phylocom comdistnt -f my.phylo.txt -s
my.sample.txt -a")
```

The result is a matrix where the community names are on the rows and columns and the values are the nearest neighbor phylogenetic dissimilarity between each set of communities. The diagonal values correspond to the MNTD values. Next we can perform a null modeling analysis for each of these metrics again by specifying the number of randomizations using the "-r" switch and the type of null model using the "-m" switch. Again we will choose 99 randomizations and the name shuffling null model "0." Lastly, we will use the "-n" switch to tell Phylocom to run a null model. If this is not included, the null model analyses will not be performed.

```
> system("./phylocom comdist -f my.phylo.txt -s
my.sample.txt -r 99 -m 0 -n")
```

We see that four matrices have been printed out to our R console. The first matrix contains the observed D_{pw} values. The second matrix contains the mean of the null distribution, the third matrix contains the standard deviation of the null distribution, and the fourth matrix contains the S.E.S. value for the D_{pw} metric again multiplied by negative one such that negative values indicate higher than expected phylogenetic beta diversity and positive values indicate lower than expected phylogenetic beta diversity. This code can be modified to use abundance using the "-a" switch and the "comdistnt" function for D_{nn}, but we will forgo those analyses for now. Also recall that you do not have to output the results to your R console. You can simply write them to your hard drive as text files using ">" followed by the new file name.

A final community phylogenetic function that we will address in Phylocom concerns whether there is more species in a community from a particular internal node in the phylogeny than expected from randomizing species names on the tips of the phylogeny. To my knowledge, this function is not yet available in R. So for the time being we can run this test in Phylocom using the "nodesigl" function.

```
> system("./phylocom nodesigl -f my.phylo.txt -s
my.sample.txt")
```

The output is a table with seven columns. The first column indicates the name of the community. The second and third columns indicate the number and name of each internal node in the phylogeny. The fourth column contains the number of taxa from that node found in that community. The fifth column indicates the median expected value and the sixth column gives the rank of the observed number of species in the null distribution of expected number of species. Note in some cases, such as the root, the null distribution contains only one possible value and therefore should not be considered. The final column contains a "+," "−," or no value to indicate the observed number of species from that node in that community is higher than expected, lower than expected, or no different from expected, respectively. This analysis can also be output as a series of Nexus phylogenies, one per community, with the node labels when there is more or less species than expected for that community. This Nexus file can be generated using the "nodesig" command and read into a program such as Mesquite for visualization.

```
> system("./phylocom nodesig -f my.phylo.txt -s
my.sample.txt > nodesig4mesquite.nex")
```

We now transition to the measurement of functional diversity using Phylocom via R. The command "comtrait" calculates the main measures of functional diversity found in Phylocom. The command requires a "-t" switch to indicate the trait file instead of the "-f" switch to indicate the phylogeny file. It also requires a "-x" switch to indicate what type of functional diversity you would like to calculate. A value of "-x 1" calculates the variance of each trait in each community, "-x 2" calculates the mean pairwise distance for each trait, "-x 3" calculates the mean nearest trait neighbor distances, and finally "-x 4" calculates the range of each trait in each community. The function also runs a null model using the same "-m" and "-r" switches we used above. To provide an example, we will calculate the variance of each trait in each community using a name shuffling null model with 99 randomizations.

```
> system("./phylocom comtrait -t my.traits.txt -s
my.sample.txt -x 1 -m 0 -r 99")
```

The output is a matrix with each row being a unique trait by community combination sorted by trait order in the traits file. The first column is the trait name, the second column is the community name, the third column is the species richness, the fourth column is the mean trait value, the fifth column is the functional diversity metric (variance in this case), the sixth column is the mean of the null distribution, the seventh column is the standard deviation of the null distribution, the eighth column is the S.E.S. for the metric, the ninth and tenth columns are the ranks, and the final column is the number or randomizations. Unfortunately, in this function the S.E.S. value is *not* multiplied by negative one further adding to the confusion, so once again some diligence is needed when considering your results.

9.3.2   Comparative Analyses in Phylocom

Although Phylocom is primarily used for the analysis of phylogenetic diversity in communities, it also has a powerful comparative methods module that can implement the calculation of phylogenetic signal and phylogenetically independent contrasts (PICs). Conveniently these measures can be calculated simultaneously using one simple command called "aot" in Phylocom.

```
> system("./phylocom aot -f my.phylo.txt -t
my.traits.txt")
```

The first output printed to your R console is the phylogenetic signal for each trait. Here phylogenetic signal is being computed as the variance of the node-level contrasts compared to a null distribution of variances generated by shuffling names of species on the phylogeny 999 times. The first column of the output is the trait name, the second column is the number of taxa analyzed, the third column is the observed variance in contrast values, and the fourth and fifth columns are the rank values of the observed variance in the null distributions. For a one-tailed test, values in the fourth column less than or equal to 50 indicate significant phylogenetic signal when using 999 randomizations.

The second component of the output reports the PIC correlation values between trait one and the other traits with the first column being the second trait being compared to the first, the second column is the correlation, the third column is the number of positive contrasts, and the final column is the number of contrasts. In order to calculate a *P*-value for the correlation, one can use a Pearson's correlation co-efficient statistical table with the degrees of freedom equaling the number of contrasts minus one.

9.3.3   Interfacing R and Phylocom for Null Modeling

In the previous two subsections, we have discussed how to run Phylocom via R, but this alone may not be a very powerful way to interface R and Phylocom. After all this approach doesn't really take advantage of R *per se*. Rather it just prevents you from having to open up a shell or terminal window and may allow you to do a few additional analyses not currently available in R. Thus we may want to quickly consider how R could be more usefully interfaced with a program like Phylocom. One potential approach that may be immediately useful is to output manipulated files from R to be used in Phylocom. In particular, we have already seen that many of the null modeling functions currently available in R are quite slow, whereas they are generally much faster in Phylocom. Further, we have seen that at least in the case of beta diversity there are no "canned" R functions for null model analyses, whereas in Phylocom there are null model analyses but they do not provide the null distribution or the estimated P-value. Here I will demonstrate a methodology for solving this particular problem, but the general approach should be useful for solving a number of tasks where one generates or modifies files in R that could be input into another program on your computer.

We being by identifying the problem and the task we want accomplished. We want to conduct a null model analysis for the phylogenetic beta diversity metric D_{pw}, where we can obtain all null values and therefore calculate a P-value. We know we can conduct a null model analysis of D_{pw} in Phylocom using the "comdist" command, but we also know that we cannot obtain the null distribution or an estimated P-value. A potential solution is to provide Phylocom 999 random phylogenies one at a time and have it calculate the D_{pw} values using each random phylogeny. We could then use all of the output files to generate null distributions and estimate P-values. For brevity, we will take this approach, but using only nine random phylogenies. We start by reading into R our original phylogeny while making sure we have our working directory set to where our Phylocom program is located.

```
> my.phylo <- read.tree("my.phylo.txt")
```

We will keep our observed community data matrix in the working directory because we will only use R to manipulate the phylogenetic tree. Specifically, we will write a simple loop that randomizes the names on the phylogeny, writes the randomized phylogeny to our working directory, and tells Phylocom to calculate the D_{pw} values using the randomized phylogeny and repeats this process nine times.

```
> for(i in 1:9){

        ## Shuffling names on the phylogeny.
        my.phylo$tip.label <- sample(my.phylo$tip.label)

        ## Writing phylogeny to working directory.
        write.tree(my.phylo, "tmp.phylogeny.txt")

        ## We use the system() command to tell Phylocom
        ## to run the comdist function using the
        ## randomizcd phylogeny tmp.phylogeny.txt. We
        ## want to output a new file for each new
        ## iteration of the randomization and we
        ## therefore need a new name for each iteration.
        ## We therefore will a paste() function that
        ## provides the command to Phylocom but changes
        ## the 'i' value in the output file name to match
        ## the 'i' in the for() loop.
        system(paste("./phylocom comdist -s my.sample.txt
        -f tmp.phylogeny.txt > ", i, ".txt", sep = ""))

}
```

The result is a series of nine files written to your working directory named "1.txt," "2.txt,"…, "9.txt." These files contain the random D_{pw} values generated for each iteration of the null model. The goal now is to read those nine files that contain matrices into R as an array with nine levels in the z-dimension. This can be accomplished by first reading in the observed output file and then combining it with the remaining files with a loop.

```
> tmp <- read.table("1.txt", sep = "\t", row.names = 1,
header = T)

> library(abind)

> for(i in 2:9){

        ## Read in the next random Dpw output
        new.tmp <- read.table(paste(i,".txt",sep = ""),
        sep = "\t", row.names = 1,
        header = T)

        ## Bind the new.tmp output to the existing array
        tmp <- abind(tmp, new.tmp, along = 3)

}
```

We now have an array with the row and column numbers equaling the number of communities and the number of levels in the z-dimension equals the number of randomizations, which in this case is 9. We can now calculate the mean and standard deviation of the null distributions using an `apply()` function.

```
> nulls.mean <- apply(tmp, c(1:2), mean, na.rm = T)

> nulls.sd <- apply(tmp, c(1:2), sd, na.rm = T)
```

These values along with the observed value can be used to calculate a S.E.S. value as described in Chap. 6, but such values can be provided by Phylocom. What are not available in Phylocom are the actual null distribution and the estimated P-value. Using the above approach that integrates R with Phylocom, we can plot the null distributions for community comparisons. For example, we can plot a histogram of the null values for the community 1 versus community 3 comparison.

```
> hist(tmp[1, 3, ]
```

We can also calculate an estimated P-value for each comparison by first calculating the observed D_{pw} values and reading them into R.

```
> system("./phylocom comdist -f my.phylo.txt -s
my.sample.txt > observed.dpw.txt")

> obs <- read.table("observed.dpw.txt", sep = "\t",
row.names = 1, header = T)
```

We add the observed results to the top layer of the array of null values using the `abind()` function.

```
> obs.nulls <- abind(obs, tmp)
```

Using this we can now calculate the estimated *P*-values as we have done before in Chap. 6.

```
> array(dim = dim(obs.nulls),
  t(apply(apply(obs.nulls,c(1, 2),rank), 3, t)))[ , , 1]
```

We now have the rank of the observed value in the null distribution from which we can estimate a *P*-value (see Chap. 6). This same approach could be used to calculate the D_{pw}', D_{nn}, and D_{nn}' metrics by using the "-a" switch in the above code calling Phylocom from R and/or using the "comdistnt" command in the above code calling Phylocom from R. As you can see, it can be fairly easy to integrate R with a program like Phylocom where you would like to sequentially feed the external program output from R. I have used a `for()` loop for this purpose, but the computational speed is not a problem because the task I am looping in R, randomizing names on a phylogeny, is a fairly simple process. Thus, when you are faced with potentially outputting a lot of files from R that you need to feed into an external program, the process can be automated using R freeing you to spend your time on other tasks.

9.4   Conclusions

In the fields of ecology and evolution we often run across interesting analytical approaches presented in an article that we think would be useful or our own research. For some that are comfortable with command line or are fluent in multiple programming languages, these published programs are easily utilized. For the majority of the potential users, though, using these programs is not an option or extremely difficult. Thankfully, the convergence of the ecological and evolutionary research communities on R has removed many of these obstacles since many of the published programs now in this literature are written in R and can be used by many researchers and this will only continue and become more common as we increasingly expect our students to become fluent in R. There are still cases, however, where the program of interest is written in another language making it difficult for a user to run or manipulate the analyses. The goal of this chapter was to demonstrate how such an obstacle can be overcome while remaining in R. I used one very popular C program, Phylocom, frequently used by ecologists for phylogenetic and functional diversity analyses and for comparative analyses, but the general concepts and approaches should apply to other programs you may be interested in written in C or other languages.

9.5   Exercises

1. Calculate the Faith's Index for your example data in Phylocom via R.
2. Noting that Phylocom does not have a randomization for Faith's Index at this time, write a phylogenetic null model shuffling species names that interfaces R with Phylocom to calculate 99 random Faith's Indices for each community and write the results of each iteration of the null model to your hard drive.
3. Take the randomization results from number 2 above, read them into R, and calculate a standardized effect size and P-value.
4. Repeat numbers 2 and 3, but this time randomize the community data matrix using an independent swap null model and do not manipulate the phylogenetic tree.
5. Simulate two trait datasets using your example phylogeny and the `fastBM()` function in the *phytools* package.
6. Write the two trait datasets in a single file to your hard drive assuring that it is formatted as required by Phylocom.
7. Via R, run the "aot" command in Phylocom using your example phylogeny and two simulated trait datasets.

References

1. Swenson, N.G. 2011. The role of evolutionary processes in producing biodiversity patterns, and the interrelationships between taxonomic, functional and phylogenetic biodiversity. *American Journal of Botany* 98: 472–480.
2. Gray, A. 1846. Analogy between the flora of Japan and that of the United States. *American Journal of Sciences and Arts* 2: 135–136.
3. Wen, J. 1999. Evolution of Eastern Asian and Eastern North American disjunct distributions in flowering plants. *Annual Review of Ecology and Systematics* 30: 421–455.
4. Latham, R.E., and R.E. Ricklefs. 1993. Continental comparisons of temperate-zone tree species diversity. In *Species diversity in ecological communities: Historical and geographical perspectives*, ed. R.E. Ricklefs and D.S. Schluter, 294–314. Chicago, IL: University of Chicago Press.
5. Guo, Q., R.E. Ricklefs, and M.L. Cody. 1998. Vascular plant diversity in eastern Asia and North America: Historical and ecological explanations. *Botanical Journal of the Linnean Society* 128: 123–136.
6. Swenson, N.G. 2013. The assembly of tropical tree communities – The advances and shortcomings of phylogenetic and functional trait analyses. *Ecography* 36: 264–276.
7. Webb, C.O., D.D. Ackerly, M.A. McPeek, and M.J. Donoghue. 2002. Phylogenies and community ecology. *Annual Review of Ecology and Systematics* 33: 475–505.
8. Vamosi, S.M., S.B. Heard, C. Vamosi, and C.O. Webb. 2009. Emerging patterns in the comparative analysis of phylogenetic community structure. *Molecular Ecology* 18: 572–592.
9. Cavender-Bares, J., K.H. Kozak, P.V.A. Fine, and P.V.A. Kembel. 2009. The merging of community ecology and phylogenetic biology. *Ecology Letters* 12: 693–715.
10. Weiher, E., and P. Keddy. 1992. Assembly rules, null models, and trait dispersion: New questions from old patterns. *Oikos* 74: 159–164.
11. Tilman, D., J. Knops, D. Wedin, P.B. Reich, M. Ritchie, and E. Siemann. 1997. The influence of functional diversity and composition on ecosystem processes. *Science* 277: 1300–1302.
12. Diaz, S., and M. Cabido. 2001. Viva la difference: Plant functional diversity matters to ecosystem processes. *Trends in Ecology and Evolution* 16: 646–655.
13. McGill, B.J., B.J. Enquist, E. Weiher, and M. Westoby. 2006. Rebuilding community ecology from functional traits. *Trends in Ecology and Evolution* 21: 178–185.
14. Webb, C.O., D.D. Ackerly, and S.W. Kembel. 2008. Phylocom: Software for the analysis of phylogenetic community structure and trait evolution. *Bioinformatics* 24: 2098–2100.
15. Paradis, E. 2011. *Analysis of phylogenetics and evolution with R*. New York: Springer.
16. Felsenstein, J. 2003. *Inferring phylogenies*. Sunderland, MA: Sinauer.
17. Harvey, P.H., and M.D. Pagel. 1991. *The comparative method in evolutionary biology*. Oxford: Oxford University Press.

N.G. Swenson, *Functional and Phylogenetic Ecology in R*, Use R!, 203
DOI 10.1007/978-1-4614-9542-0, © Springer Science+Business Media New York 2014

18. Brooks, D.R., and D.A. McLennan. 1991. *Phylogeny, ecology, and behavior: A research program in comparative biology*. Chicago, IL: University of Chicago Press.

19. Borcard, D., F. Gillet, and P. Legendre. 2011. *Numerical ecology in R*. New York: Springer.

20. Maddison, W.P., and D.R. Maddison. 2011. Mesquite: A modular system for evolutionary analysis. Version 2.75. http://mesquiteproject.org

21. Losos, J.B. 1994. An approach to the analysis of comparative data when a phylogeny is unavailable or incomplete. *Systematic Biology* 43: 117–123.

22. Kraft, N.J.B., W.K. Cornwell, C.O. Webb, and D.D. Ackerly. 2007. Trait evolution, community assembly, and the phylogenetic structure of ecological communities. *The American Naturalist* 170: 271–283.

23. Brock, C.D., L.J. Harmon, and M.E. Alfaro. 2011. Testing for temporal variation in diversification rates when sampling is incomplete and nonrandom. *Systematic Biology* 60: 410–419.

24. Revell, L.M., L.J. Harmon, and D.C. Collar. 2008. Phylogenetic signal, evolutionary process, and rate. *Systematic Biology* 57: 591–601.

25. Faith, D.P. 1992. Conservation evaluation and phylogenetic diversity. *Biological Conservation* 61: 1–10.

26. Faith, D.P. 1994. Genetic diversity and taxonomic priorities for conservation. *Biological Conservation* 68: 69–74.

27. Faith, D.P. 2002. Quantifying biodiversity: A phylogenetic perspective. *Conseration Biology* 16: 248–252.

28. Webb, C.O. 2000. Exploring the phylogenetic structure of ecological communities: An example for rainforest trees. *The American Naturalist* 156: 145–155.

29. Tofts, R., and J. Silvertown. 2000. A phylogenetic approach to community assembly from a local species pool. *Proceedings of the Royal Society B* 267: 363–369.

30. Cavender-Bares, J., D.D. Ackerly, D.A. Baum, and F.A. Bazzaz. 2004. Phylogenetic overdispersion in Floridian oak communities. *The American Naturalist* 163: 823–843.

31. Cavender-Bares, J., A. Keen, and B. Miles. 2006. Phylogenetic structure of Floridian plant communities depends on taxonomic and spatial scale. *Ecology* 87(Supplement): S109–S122.

32. Swenson, N.G., B.J. Enquist, J. Pither, J. Thompson, and J.K. Zimmerman. 2006. The problem and promise of scale dependency in community phylogenetics. *Ecology* 87: 2418–2424.

33. Swenson, N.G., B.J. Enquist, J. Thompson, and J.K. Zimmerman. 2007. The influence of spatial and size scales on phylogenetic relatedness in tropical forest communities. *Ecology* 88: 1770–1780.

34. Gillespie, R.G. 2004. Community assembly through adaptive radiation in Hawaiian spiders. *Science* 303: 356–359.

35. Losos, J.B. 1992. The evolution of convergent structure in Caribbean *Anolis* communities. *Systematic Biology* 41: 403–420.

36. Verdu, M., and G. Pausas. 2007. Fire drives phylogenetic clustering in Mediterranean Basin woody plant communities. *Journal of Ecology* 95: 1316–1323.

37. Cardillo, M., J.L. Gittleman, and A. Purvis. 2008. Global patterns in the phylogenetic structure of island mammal assemblages. *Proceedings of the Royal Society B* 275: 1549–1556.

38. Anderson, T.M., M.A. Lachance, and W.T. Starmer. 2004. The relationship of phylogeny to community structure: The cactus yeast community. *The American Naturalist* 164: 709–721.

39. MacArthur, R., and R. Levins. 1967. The limiting similarity, convergence, and divergence of coexisting species. *The American Naturalist* 101: 377–385.

40. Keddy, P.A. 1992. Assembly and response rules: Two goals for predictive community ecology. *Journal of Vegetation Science* 3: 157–164.

41. Hubbell, S.P. 1979. Tree dispersion, abundance, and diversity in a tropical dry forest. *Science* 203: 1299–1309.

42. Hubbell, S.P., and R.B. Foster. 1986. Commonness and rarity in a Neotropical forest: Implications for tropical tree conservation. In *Conservation biology: The science of scarcity and diversity*, ed. M.E. Soule, 205–231. Sunderland, MA: Sinauer.

43. Hubbell, S.P. 2001. *The unified neutral theory of biodiversity and biogeography*. Princeton, NJ: Princeton University Press.
44. Mayfield, M.M., and J.M. Levine. 2010. Opposing effects of competitive exclusion on the phylogenetic structure of communities. *Ecology Letters* 13: 1085–1093.
45. Jaccard, P. 1926. Le coefficient generique et le coefficient communaute dans la flore marocaine. *Bulletin de la Societe Vaudoise des Sciences Naurelles* 2: 385–403.
46. Jaccard, P. 1940. Coefficient generique reel et coefficient generique probable. *Bulletin de la Societe Vaudoise des Sciences Naurelles* 61: 117–136.
47. Maillefer, A. 1928. Les courbes de Willis: Repartition des especes dans les genres de differente etendue. *Bulletin de la Societe Vaudoise des Sciences Naurelles* 56: 617–631.
48. Elton, C. 1946. Competition and the structure of ecological communities. *Journal of Animal Ecology* 15: 54–68.
49. Simberloff, D.S. 1970. Taxonomic diversity of island biotas. *Evolution* 24: 23–47.
50. Grant, P.R., and I. Abbott. 1980. Interspecific competition, island biogeography and null hypotheses. *Evolution* 34: 332–341.
51. Harvey, P.H., R.K. Colwell, J.W. Silvertown, and R.M. May. 1983. Null models in ecology. *Annual Review of Ecology and Systematics* 14: 189–211.
52. Jarvinen, O. 1982. Species-to-genus ratios in biogeography: A historical note. *Journal of Biogeography* 9: 363–370.
53. Wiens, J.J., and C.H. Graham. 2005. Niche conservatism: Integrating evolution, ecology and conservation biology. *Annual Review of Ecology and Systematics* 36: 519–539.
54. Peterson, A.T., J. Soberon, and V. Sanchez-Cordero. 1999. Conservatism of ecological niches in evolutionary time. *Science* 285: 1265–1267.
55. Wiens, J.J., D.D. Ackerly, A.P. Allen, B.L. Anacker, L.B. Buckley, H.V. Cornell, E.I. Damschen, T.J. Davies, J.A. Grytnes, S.P. Harrison, B.A. Hawkins, R.D. Holt, C.M. McCain, and P.R. Stephens. 2010. Niche conservatism as an emerging principle in ecology and conservation biology. *Ecology Letters* 13: 1310–1314.
56. Webb, C.O., and M.J. Donoghue. 2005. Phylomatic: Tree assembly for applied phylogenetics. *Molecular Ecology Notes* 5: 181–183.
57. Webb, C.O., C.H. Cannon, and S.J. Davies. 2008. Ecological organization, biogeography, and the phylogenetic structure of tropical forest tree communities. In *Tropical forest community ecology*, ed. W.P. Carson and S.S.A. Schnitzer, 79–97. Oxford: Blackwell.
58. Kembel, S.W., P.D. Cowan, M.R. Helmus, W.K. Cornwell, H. Morlon, D.D. Ackerly, S.P. Blomberg, and C.O. Webb. 2010. Picante: R tools for integrating phylogenies and ecology. *Bioinformatics* 26: 1463–1464.
59. Vane-Wright, R.I., C.J. Humphries, and P.H. Williams. 1991. What to protect – Systematics and the agony of choice. *Biological Conservation* 55: 235–254.
60. Harmon, L.J., J.T. Weir, C.D. Brock, R.E. Glor, and W. Challenger. 2008. GEIGER: Investigating evolutionary radiations. *Bioinformatics* 24: 129–131.
61. Mooers, A.O., S.B. Heard, and E. Chrostowski. 2005. Evolutionary heritage as a metric for conservation. In *Phylogeny and conservation*, ed. A. Purvis, T.L. Brooks, and J.L. Gittleman, 120–138. Oxford: Oxford University Press.
62. Barker, G.M. 2002. Phylogenetic diversity: A quantitative framework for measurement of priority and achievement in biodiversity conservation. *Biological Journal of the Linnean Society* 76: 165–194.
63. Kembel, S.W., and S.P. Hubbell. 2006. The phylogenetic structure of a neotropical forest tree community. *Ecology* 87(Supplement): S86–S99.
64. Swenson, N.G., D.L. Erickson, X. Mi, N.A. Bourge, J. Forero-Montana, X. Ge, R. Howe, J.K. Lake, X. Liu, K. Ma, N. Pei, J. Thompson, M. Uriarte, A. Wolf, S.J. Wright, W. Ye, J. Zhang, J.K. Zimmerman, and W.J. Kress. 2012. Phylogenetic and functional alpha and beta diversity in temperate and tropical tree communities. *Ecology* 93: S112–S125.
65. Yang, J., X. Ci, G. Zhang, N.G. Swenson, L. Sha, C.C. Baskin, J. Li, M. Cao, J.W.F. Slik, and L. Lin. in press. Functional and phylogenetic assembly in a Chinese tropical tree community across size classes, spatial scales and habitats. *Functional Ecology*.

66. Mi, X., N.G. Swenson, R. Valencia, W.J. Kress, D.L. Erickson, A. Perez-Castaneda, H. Ren, S.H. Su, N. Gunatilleke, S. Gunatilleke, Z. Hao, W. Ye, M. Cao, H.S. Suresh, H.S. Dattaraj, R. Sukumar, and K. Ma. 2012. The contribution of rare species to community phylogenetic diversity across a global network of forest plots. *The American Naturalist* 180: E17–E30.

67. Webb, C.O., and N.C.A. Pitman. 2002. Phylogenetic balance and ecological evenness. *Systematic Biology* 51: 898–907.

68. Rao, C.R. 1982. Diversity and dissimilarity coefficients: A unified approach. *Theoretical Population Biology* 21: 24–43.

69. Hardy, O.J., and B. Senterre. 2007. Characterizing the phylogenetic structure of communities by additive partitioning of phylogenetic diversity. *Journal of Ecology* 95: 493–506.

70. Helmus, M.R., T.J. Bland, C.K. Williams, and A.R. Ives. 2007. Phylogenetic measures of biodiversity. *The American Naturalist* 169: E68–E83.

71. Vellend, M., W.K. Cornwell, K. Magnuson-Ford, and A.O. Mooers. 2010. Measuring phylogenetic biodiversity. In *Biological diversity: Frontiers in measurement and assessment*, ed. A.E. Magurran and B.J. McGill, 193–206. Oxford: Oxford University Press.

72. Brown, W.L., and E.O. Wilson. 1956. Character displacement. *Systematic Zoology* 5: 49–65.

73. Weiher, E., G.D.P. Clarke, and P.A. Keddy. 1998. Community assembly rules, morphological dispersion, and the coexistence of plant species. *Oikos* 81: 309–322.

74. Thompson, K., S.H. Hillier, J.P. Grime, C.C. Bossard, and S.R. Band. 1996. A functional analysis of a limestone grassland community. *Journal of Vegetation Science* 7: 371–380.

75. Grime, J.P. 2006. Trait convergence and trait divergence in herbaceous plant communities: Mechanisms and consequences. *Journal of Vegetation Science* 17: 255–260.

76. Petchey, O.L., and K.J. Gaston. 2002. Functional diversity (FD), species richness, and community composition. *Ecology Letters* 5: 402–411.

77. Kraft, N.J.B., R. Valencia, and D.D. Ackerly. 2008. Functional traits and niche-based tree community assembly in an Amazonian forest. *Science* 322: 580–582.

78. Swenson, N.G., and B.J. Enquist. 2007. Ecological and evolutionary determinants of a key plant functional trait: Wood density and its community-wide variation across latitude and elevation. *American Journal of Botany* 91: 451–459.

79. Swenson, N.G., and B.J. Enquist. 2009. Opposing assembly mechanisms in a Neotropical dry forest: Implications for phylogenetic and functional community ecology. *Ecology* 90: 2161–2170.

80. Swenson, N.G., P. Anglada-Cordero, and J.A. Barone. 2011. Deterministic tropical tree community turnover: Evidence from patterns of functional beta diversity along an elevational gradient. *Proceedings of the Royal Society B* 278: 877–884.

81. Liu, X., N.G. Swenson, S.J. Wright, L. Zhang, K. Song, Y. Du, J. Zhang, X. Mi, and K. Ma. 2012. Covariation in plant functional traits and soil fertility within two species-rich forests. *PLoS One* 7: e34767.

82. Liu, X., N.G. Swenson, J. Zhang, and K. Ma. 2013. The environment and space, not phylogeny, determine trait dispersion in a subtropical forest. *Functional Ecology* 27: 264–272.

83. Jung, V., C. Violle, C. Mondy, L. Hoffmann, and S. Muller. 2010. Intraspecific variability and trait-based community assembly. *Journal of Ecology* 98: 1134–1140.

84. Shipley, B., D. Vile, and E. Garnier. 2006. From plant traits to plant communities: A statistical mechanistic approach to biodiversity. *Science* 314: 812–814.

85. Laughlin, D.C., P.Z. Fule, D.W. Huffman, J. Crouse, and E. Laliberte. 2011. Climatic constraints on trait-based forest assembly. *Journal of Ecology* 99: 1489–1499.

86. Baraloto, C., O.J. Hardy, C.E.T. Paine, K.G. Dexter, C. Cruaud, L.T. Dunning, M.A. Gonzalez, J.F. Molino, D. Sabatier, V. Savolainen, and J. Chave. 2012. Using functional traits and phylogenetic trees to examine the assembly of tropical tree communities. *Journal of Ecology* 100: 690–701.

87. Paine, C.E.T., C. Baraloto, J. Chave, and B. Herault. 2011. Functional traits of individual trees reveal ecological constraints on community assembly in tropical rain forests. *Oikos* 120: 720–727.

88. Siefert, A. 2012. Spatial patterns of functional divergence in old-field plant communities. *Oikos* 121: 907–914.

89. Siefert, A., C. Ravenscroft, M.D. Weiser, and N.G. Swenson. 2013. Functional beta diversity patterns reveal deterministic community assembly processes in eastern North American trees. *Global Ecology and Biogeography* 22: 682–691.

90. Cornwell, W.K., D.W. Schwilk, and D.D. Ackerly. 2006. A trait-based test for habitat filtering: Convex hull volume. *Ecology* 87: 1465–1471.

91. Cadotte, M.W., B.J. Cardinale, and T.H. Oakley. 2008. Evolutionary history and the effect of biodiversity on plant productivity. *Proceedings of the National Academy of Sciences of the United States of America* 105: 17012–17017.

92. Laliberte, E., and P. Legendre. 2010. A distance-based framework for measuring functional diversity from multiple traits. *Ecology* 91: 299–305.

93. Ricklefs, R.E., and K. O'Rourke. 1975. Aspect diversity in moths: A temperate-tropical comparison. *Evolution* 29: 313–324.

94. Ricklefs, R.E., and J. Travis. 1980. A morphological approach to the study of avian community organization. *Auk* 97: 321–338.

95. Shepherd, U.L. 1998. A comparison of species diversity and morphological diversity across the North American latitudinal gradient. *Journal of Biogeography* 25: 19–29.

96. Stevens, R.D., S.B. Cox, M.R. Willig, and R.E. Strauss. 2003. Patterns of functional diversity across an extensive environmental gradient: Vertebrate consumers, hidden treatments, and latitudinal trends. *Ecology Letters* 6: 1099–1108.

97. Swenson, N.G., B.J. Enquist, J. Pither, A.J. Kerkhoff, B. Boyle, M.D. Weiser, J.J. Elser, W.F. Fagan, J. Forero-Montana, N. Fyllas, N.J.B. Kraft, J.K. Lake, A.T. Moles, S. Patino, O.L. Phillips, C.A. Price, P.B. Reich, C.A. Quesada, J.C. Stegen, R. Valencia, I.J. Wright, S.J. Wright, S. Andelman, P.M. Jorgensen, T.E. Lacher Jr., A. Monteagudo, P. Nunez-Vargas, R. Vasquez, and K.M. Nolting. 2012. The biogeography and filtering of woody plant functional diversity in North and South America. *Global Ecology and Biogeography* 21: 798–808.

98. Cadotte, M.W., J. Cavender-Bares, D. Tilman, and T.H. Oakley. 2009. Using phylogenetic, functional and trait diversity to understand patterns of plant community productivity. *PLoS One* 4: e5695.

99. Westoby, M. 1998. A leaf-height-seed (LHS) plant ecology strategy scheme. *Plant and Soil* 199: 213–227.

100. Weiher, E., A. Van der Werf, K. Thompson, M. Roderick, E. Garnier, and O. Eriksson. 1999. Challenging Theophrastus: A common core list of plant traits for functional ecology. *Journal of Vegetation Science* 10: 609–620.

101. Westoby, M., D.S. Falster, A.T. Moles, P.A. Vesk, and I.J. Wright. 2002. Plant ecological strategies: Some leading dimensions of variation between species. *Annual Review of Ecology and Systematics* 33: 125–159.

102. Westoby, M., and I.J. Wright. 2006. Land-plant ecology on the basis of functional traits. *Trends in Ecology and Evolution* 21: 261–268.

103. Swenson, N.G., and M.D. Weiser. 2010. Plant geography upon the basis of functional traits: An example from eastern North American trees. *Ecology* 91: 2234–2241.

104. Gower, J.C. 1971. A general coefficient of similarity and some of its properties. *Biometrics* 27: 4979–4983.

105. Tuomisto, H., and K. Ruokolainen. 2006. Analyzing of explaining beta diversity? Understanding the targets of different methods of analysis. *Ecology* 87: 2697–2708.

106. Legendre, P., D. Borcard, and P.R. Peres-Neto. 2008. Analyzing of explaining beta diversity? Comment. *Ecology* 89: 3238–3244.

107. Tuomisto, H., and K. Ruokolainen. 2008. Analyzing or explaining beta diversity? Reply. *Ecology* 89: 3244–3256.

108. Anderson, M.J., T.O. Crist, J.M. Chase, M. Vellend, B.D. Inouye, A.L. Freestone, N.J. Sanders, H.V. Cornell, L.S. Comita, K.F. Davies, S.P. Harrison, N.J.B. Kraft, J.C. Stegen, and N.G. Swenson. 2011. Navigating the multiple meanings of beta diversity: A roadmap for the practicing ecologist. *Ecology Letters* 14: 19–28.

109. Fukami, T., T.M. Bezemer, S.R. Mortimer, and W.H. van der Putten. 2005. Species divergence and trait convergence in experimental plant community assembly. *Ecology Letters* 8: 1283–1290.
110. Swenson, N.G., J.C. Stegen, S.J. Davies, D.L. Erickson, J. Forero-Montana, A.H. Hurlbert, W.J. Kress, J. Thompson, M. Uriarte, S.J. Wright, and J.K. Zimmerman. 2012. Temporal turnover in the composition of tropical tree communities: Functional determinism and phylogenetic stochasticity. *Ecology* 93: 490–499.
111. Zhang, J., N.G. Swenson, S. Chen, X. Liu, Z. Li, J. Huang, X. Mi, and K. Ma. 2013. Phylogenetic beta diversity in tropical forests: Implications for the role of geographical and environmental distance. *Journal of Systematics and Evolution* 51: 71–85.
112. Swenson, N.G. 2011. Phylogenetic beta diversity metrics, trait evolution and inferring the functional beta diversity of communities. *PLoS One* 6: e21264.
113. Martin, A.P. 2002. Phylogenetic approaches for describing and comparing the diversity of microbial communities. *Applied and Environmental Microbiology* 68: 3673–3682.
114. Lozupone, C., and R. Knight. 2005. UniFrac: A new phylogenetic method for comparing microbial communities. *Applied and Environmental Microbiology* 71: 8228–8235.
115. Graham, C.H., and P.V.A. Fine. 2008. Phylogenetic beta diversity: Linking ecological and evolutionary processes across space and time. *Ecology Letters* 11: 1265–1277.
116. Lozupone, C.A., M. Hamady, S.T. Kelley, and R. Knight. 2007. Quantitative and qualitative beta diversity measures lead to different insights into factors that structure microbial communities. *Applied Environmental Microbiology* 73: 1576–1585.
117. Chen, J., K. Bittinger, E.S. Charlson, C. Hoffmann, J. Lewis, G.D. Wu, R.G. Collman, F.D. Bushman, and H. Li. 2012. Associating microbiome composition with environmental covariates using generalized UniFrac distances. *Bioinformatics* 28: 2106–2113.
118. Chang, Q., Y. Luan, and F. Sun. 2011. Variance adjusted weighted UniFrac: A powerful beta diversity measure for comparing communities based on phylogeny. *BMC Bioinformatics* 12: 118.
119. Bryant, J.B., C. Lamanna, H. Morlon, A.J. Kerkhoff, B.J. Enquist, and J.L. Green. 2008. Microbes on mountainsides: Contrasting elevational patterns of bacterial and plant diversity. *Proceedings of the National Academy of Sciences of the United States of America* 105: 7774–7778.
120. Ricotta, C., and S. Burrascano. 2008. Beta diversity for functional ecology. *Preslia* 80: 61–71.
121. Ricotta, C., and S. Burrascano. 2009. Testing for differences in beta diversity with asymmetric dissimilarities. *Ecological Indicators* 9: 719–724.
122. Gotelli, N.J., and G.R. Graves. 1996. *Null models in ecology*. Washington, DC: Smithsonian Books.
123. Colwell, R.K., and D.W. Winkler. 1984. A null model for null models in biogeography. In *Ecological communities: Conceptual issues and the evidence*, ed. D.R. Strong Jr., D. Simberloff, L.G. Abele, and A.B. Thistle, 344–359. Princeton, NJ: Princeton University Press.
124. Kraft, N.J.B., and D.D. Ackerly. 2010. Functional trait and phylogenetic tests of community assembly across spatial scales in an Amazonian forest. *Ecological Monographs* 80: 401–422.
125. Gotelli, N.J., and G.L. Entsminger. 2001. Swap and fill algorithms in null model analysis: Rethinking the knight's tour. *Oecologia* 129: 281–291.
126. Hardy, O.J. 2008. Testing the spatial phylogenetic structure of local communities: Statistical performances of different null models and test statistics on a locally neutral community. *Journal of Ecology* 96: 914–926.
127. Lake, J.K., and A. Ostling. 2009. Comment on 'Functional traits and niche-based community assembly in an Amazonian forest'. *Science* 324: 1015.
128. Kraft, N.J.B., and D.D. Ackerly. 2009. Response to comment on 'Functional traits and niche-based community assembly in an Amazonian forest'. *Science* 324: 1015.
129. Felsenstein, J. 1985. Phylogenies and the comparative method. *The American Naturalist* 125: 1–15.

130. Grafen, A. 1989. The phylogenetic regression. *Philosophical Transactions of the Royal Society B* 326: 119–156.
131. Grafen, A. 1992. The uniqueness of the phylogenetic regression. *Journal of Theoretical Biology* 156: 405–523.
132. Martins, E.P., and T.F. Hansen. 1997. Phylogenies and the comparative method: A general approach to incorporating phylogenetic information into the analysis of interspecific data. *The American Naturalist* 149: 646–667.
133. Diniz-Filho, J.A.F., C.E. Romos de Sant'Ana, and L.M. Bini. 1998. An eigenvector method for estimating phylogenetic inertia. *Evolution* 52: 1247–1262.
134. Freckleton, R.P., N. Cooper, and W. Jetz. 2011. Comparative methods as a statistical fix: The dangers of ignoring an evolutionary model. *The American Naturalist* 178: E10–E17.
135. Blomberg, S.P., and T. Garland Jr. 2002. Tempo and mode in evolution: Phylogenetic inertia, adaptation and comparative methods. *Journal of Evolutionary Biology* 15: 899–910.
136. Losos, J.B. 2008. Phylogenetic niche conservatism, phylogenetic signal and the relationship between phylogenetic relatedness and ecological similarity among species. *Ecology Letters* 11: 995–1007.
137. Wiens, J.J. 2008. Commentary: Niche conservatism déjà vu. *Ecology Letters* 11: 1004–1005.
138. Harmon, L.J., and R.E. Glor. 2010. Poor statistical performance of the Mantel Test in phylogenetic comparative analyses. *Evolution* 64: 2173–2178.
139. Blomberg, S.P., T. Garland Jr., and A.R. Ives. 2003. Testing for phylogenetic signal in comparative data: Behavioral traits are more labile. *Evolution* 57: 717–745.
140. Garland Jr., T., P.E. Midford, and A.R. Ives. 1999. An introduction to phylogenetically based statistical methods, with a new method for confidence intervals on ancestral values. *American Zoologist* 39: 374–388.
141. Pagel, M. 1999. Inferring the historical patterns of biological evolution. *Nature* 401: 877–884.
142. Harmon, L.J., J.A. Schulte II, A. Larson, and J.B. Losos. 2003. Tempo and mode of evolutionary radiation in Iguanian lizards. *Science* 301: 961–964.
143. Mayfield, M.M., M.F. Boni, and D.D. Ackerly. 2009. Traits, habitats, and clades: Identifying traits of potential importance to environmental filtering. *The American Naturalist* 174: E1–E22.
144. Pavoine, S., E. Vela, S. Gachet, G. de Belair, and M.B. Bonsall. 2011. Linking patterns in phylogeny, traits, abiotic variables and space: A novel approach to linking environmental filtering and plant community assembly. *Journal of Ecology* 99: 165–175.
145. Greenacre, M.J. 1984. *Theory and applications of correspondence analysis.* London: Academic.

Index

N.G. Swenson, *Functional and Phylogenetic Ecology in R*, Use R!,
DOI 10.1007/978-1-4614-9542-0, © Springer Science+Business Media New York 2014

MIX
Papier aus verantwortungsvollen Quellen
Paper from responsible sources
FSC® C105338
www.fsc.org

If you have any concerns about our products,
you can contact us on
ProductSafety@springernature.com

In case Publisher is established outside the EU,
the EU authorized representative is:
Springer Nature Customer Service Center GmbH
Europaplatz 3, 69115 Heidelberg, Germany

Printed by Libri Plureos GmbH
in Hamburg, Germany